Annette Vaillancourt

D1264136

MYTHS OF GENDER

MYTHS OF GENDER

BIOLOGICAL THEORIES

ABOUT WOMEN AND MEN

ANNE FAUSTO-STERLING

Basic Books, Inc., Publishers   New York

Library of Congress Cataloging-in-Publication Data

Fausto-Sterling, Anne, 1944–
  Myths of gender.

  Bibliographic references: p. 225
  Includes index.
  1. Sex differences. 2. Feminism. 3. Sexism.
4. Human biology—Philosophy. 5. Prejudices.
I. Title.
QP81.5.F38 1985   155.33   85–47561
ISBN 0–465–04790–4

Copyright © 1985 by Basic Books, Inc.
Printed in the United States of America
Designed by Vincent Torre
85 86 87 88 HC 9 8 7 6 5 4 3 2 1

For my parents,

Philip and Dorothy Sterling,

who have shared with me

their visions of the world,

including its past shames and glories,

its present pains and wonders,

and its future community

CONTENTS

ACKNOWLEDGMENTS

THIS PROJECT has taken more than four years to complete. During that time many people have supported the work in many different ways. At the most tangible level, I have received fellowships from the Wellesley Center for Research on Women at Wellesley College and from the Pembroke Center for Teaching and Research on Women at Brown University. The Brown University administration has cooperated in granting me both a sabbatical leave and leaves of absence to utilize these fellowships. Other forms of practical help have come from the librarians at both the Sciences and Rockefeller libraries at Brown University, who went out of their way to deal with my sometimes urgent and often oddball requests. I have deeply appreciated their professional competence and their personal interest in my work. Christina Crosby and Carol Cohen both did extensive library work for me at critical times in the book's development, and Carol King typed more than one version of the manuscript with great skill and, perhaps most importantly, with good humor. Her cheerfulness and sense of steadiness often provided balance for my more mercurial self-presentation. Lily Hsieh kept other parts of my professional life going, enabling me to focus on the book.

I have had the extraordinary good fortune to belong to two different extended families, known informally as the Tacky Family and the Alstead Book Society (and Eating Club). The gifts of their individual and collective friendships have on occasion kept me from drowning and have, among many other things, freed me to work on a project so seemingly unending as this book. To two Barbaras, David, Bill, Marlene, Nelson, Harold, Susan, two Elizabeths, Christina, Don, Joan, and Karen: thanks and may our friendship, love, and loyalty sustain us throughout our lives. A special thanks goes to the children of these families for sharing with me their unique and wonderful visions of the world. Daniel, Michael, Luke, Karl, Nathan, Toby, Bryony, and Liz: as you descend into the abyss of adulthood, may you never lose sight of the freshness and openness of childhood.

Many people have offered very particular kinds of help. Karen Romer believed in my ability to write a book long before I did. She constantly encouraged me, finding in the midst of her own busy life the time to read and comment on two manuscript drafts. Joan Scott fulfilled her professional duties as a literary *schotkchun* by putting me in contact with my editor-to-be, Steve Fraser. In writing this book I was kept fully in touch with the popular press through the devotion of several avid media watchers. My husband, Nelson Fausto, my aunt, Alice Lake, and my mother, Dorothy Sterling, as well as quite a number of students and librarians at Brown University kept me papered with clippings and alerted to the latest relevant magazine articles. Lenny Erickson kept me abreast of the *New York Review of Books* and lent me his copy whenever necessary. Without their interest and consistent help my job would have been much harder. My father, Philip Sterling, lent his especially keen editorial eye to major parts of the manuscript, and his interest and suggestions have helped considerably in the final shaping of the work. So, too, have the excellent skills of my editor at Basic Books, Steven Fraser. Discussions with him helped put the book in final shape. His prompt and considerate feedback and his quick eye for what does and does not work moved things along.

Ruth Bleier and I exchanged chapters as she worked simultaneously on her book *Science and Gender* (Pergamon, 1984). Ruth generously shared both her references and her encouragement. Thanks, too, to the members of the 1982 Pembroke Seminar for their comments and to Harriet Whitehead for repeatedly forcing me to think through my assumptions just one more time. The following people have also taken their valuable time to read and comment on various drafts of chapters: Joan Scott, Nelson Fausto, Dorothy Sterling, Elizabeth Kirk, Barbara Johnson, Peter Sterling, Susan P. Benson, Diana Jackson, Christina Crosby, Elizabeth Weed, Annette Coleman, Sandy Parra, Jennifer Zollner, Louise Lamphere, Phil Quinn, James McIlwain, Peter Heywood, Barbara Melosh, Michelle Wittig, Margaret Kidwell, Richard Lewontin, Carole Offir, Pat Blough, Julia Sherman, Jane Hitti, Natalie Kampen, and Barrie Thorne.

It is, I believe, traditional to absolve everyone who has helped me from responsibility for errors in the final book, and of course I

ACKNOWLEDGMENTS

do that gladly, knowing that what is in the book—for better or for worse—results ultimately from my own decision making. Whatever strengths may be found, however, come from the collective process suggested in the acknowledgments and from the help and friendship of many others not specifically cited in these pages.

MYTHS OF GENDER

1

THE BIOLOGICAL
CONNECTION:
AN INTRODUCTION

> As Darwin demonstrated . . . we males have been
> born the fittest for three billion years. From that
> constantly on-the-make little treemouse (the
> lemur . . .) to Mailer the magnificent, the DNA of
> the male Y chromosome has programmed us to lead
> our sisters.   —EDGAR BERMAN
> *The Compleat Chauvinist*, 1982

THESE are difficult times. The middle-class family of the
1950s, headed by a husband, managed by a wife, and enlivened by
two children and a collie, is no more. In its stead we have single-
parent families, contract marriages, palimony suits, and serial monogamy. Women demand equal pay for work of equal value and
take assertiveness training, and feminists organize to change long-
standing political and social structures. Men are unsure about
whether to expect a thank you or a snarl when they hold a door
open for a woman, and clothing designers offer us the world of
unisex dress. Hair length doesn't matter; the gender gap in finishing
marathon races gets smaller every year; and "affirmative action"
has become embedded in our language. The variety and rapidity of
the changes symbolized by these random examples have generated
an expressly political backlash by the "New Right," doubts about
individual identity, and fears of sexual obliteration-through-equality.
In response to such personal and social upheaval, professionals
throughout the country—scientists, journalists, economists, and

3

politicians—have begun to search out the real truths about sexual differences. And therein lies a tale.

Over the years physicians, biologists, and anthropologists have had a lot to say about women's place in the world. In the nineteenth century, some scientists wrote that women who work to obtain economic independence set themselves up for "a struggle against Nature,"[1] while author after author used Darwin's theory of evolution to argue that giving the vote to women was, evolutionarily speaking, retrogressive.[2] Physicians and educators alike warned that young women who engage in long, hard hours of study will badly damage their reproductive systems, perhaps going insane to boot. With these warnings came grim predictions about the end of the (white, middle-class) human race.[3] Ironically, feminists of the period often used the same biological arguments to support their own points of view.[4] Antoinette Brown Blackwell, for example, garnered evidence from the natural world to prove the fundamental equality of men and women,[5] and Eliza Gamble argued outright for the natural superiority of the female sex.[6]

Today, too, many scientists respond to the issue of sexual equality and the social and political upheaval that has accompanied it by offering us their insights, suggesting in all sincerity that, however well intentioned, the women's liberation movement and its fellow travelers want biologically unnatural changes that would bring grief to the human race. Sociobiologists, for example, suggest that our evolutionary history deeply affects our most intimate personal relationships. Man's natural sexuality sends him in search of many sex partners, making him an unstable mate at best, while woman's biological origins destine her to keep the home fires burning, impelling her to employ trickery and deceit to keep hubby from straying. The battle between the sexes is ancient; its origin lies hidden deep in our genes.[7] One sociobiologist even argues that male and female sexuality are so different, so at odds, that it makes sense to think of the two sexes as separate species.[8] In general these scientists emphasize difference, the biological logic of male-female conflict, and the dim prospects for change.

Even physical violence fits into the picture. Dr. Katharina Dalton, a British physician who has made a name for herself by publicizing and then inventing treatments for something she calls the Premenstrual Syndrome, suggests that monthly hormone fluctuations may cause afflicted women to unknowingly injure themselves, claiming afterward that their husbands had beaten them:

"All too often the patient herself is not fully aware of the distress caused by her periodic tantrums. . . . When a woman demonstrates bruises as signs of her husband's cruelty it is well to remember the possibility that these may be spontaneous bruises of the premenstruum."[9]

Biologically based argumentation has even invaded the criminal court system. In the trial of a woman who had used an automobile to run over and kill her boyfriend, Dalton testified that the woman had suffered from premenstrual derangement and should not be held legally responsible for her acts. As a result the woman received a conditional discharge from jail.[10] A similar defense is now valid in French courts,[11] while in the United States a judge recently acquitted a dentist accused of rape and sodomy, after the defendant claimed that his girlfriend had filed the charges during a period of premenstrual irrationality.[12] Whether the idea that we are mere agents of our own bodies will make deep inroads into the criminal justice system remains to be seen. If this idea does take hold, the erosion of personal responsibility for one's actions—be it a woman who commits murder or a man who batters or rapes—would be an inevitable consequence.

Lost also would be our society's ability to recognize large-scale violence as a social problem to be dealt with in a public arena. One recent study, for instance, suggests that a woman in a large city stands a 26 percent chance of being raped during her lifetime. That statistic increases to 50 percent when the possibility of rape attempts is taken into account.[13] Are we to believe that these statistics result from the male's ungovernable mating urge combined with the false reporting of distraught premenstrual women? Or must we face up to the conclusion that sexual violence is somehow embedded in the social fabric? If we believe the former, then there's not much to be done about it; if we believe the latter, we must collectively endeavor to change the assumptions and attitudes of our culture—a complex and difficult task. Clearly, what we think about the biological basis of criminal violence matters a great deal.

So, too, do our beliefs about women's economic welfare. American women grow poorer every year. Currently two out of every three impoverished adults are women.[14] Some estimates hold that by the year 2000, 90 percent of all Americans living below the poverty line will be older women and young women with dependent children. The causes of this increasing "feminization of poverty" (as the problem has been dubbed) are complex, but chief among them is the fact that women earn only fifty-nine cents for every dollar paid to their male counterparts. Explanations for this statistic include both pay discrimination on the job and prejudices that relegate women to employment in only the lowest-paying job categories. At least those are among the reasons given by both women's rights organizations *and* the President's Commission on Civil Rights. But conservative writer George Gilder has another suggestion, something he calls "the biological factor." He argues that men are by nature more aggressive than women, and that this heightened aggressiveness, men's larger physical size, and what he calls "the male need to dominate" combine to make men natural group leaders. Furthermore, in prehistoric times males hunted to provide food, an evolutionary history that makes the desire to provide one of "the deepest instincts of men." Women—who, according to Gilder's version of prehistory, stayed near the hearth tending the kids and waiting for the meat to arrive—continue to the present day to want nothing more than to stay at home. Thus, when women do work, they simply cannot give it the same all-out effort offered by men. "These differences between the sexes," he writes, "fully explain all gaps in earning."[15]

From this line of reasoning, it would follow that employment discrimination, although it may exist from time to time, has little to do with female poverty. Philosophy professor Michael Levin carries this logic one step further in his argument that affirmative action to promote equal employment and increase female earning power is both inappropriate and doomed to fail. Why? Because women really *are* biologically inferior. To Levin's eye the biological facts dictate that women cannot compete. The price for equal employment can only be lower quality work.[16]

U.S. Bureau of Labor Statistics projections for job prospects in the coming decade reveal several trends that concern women. In the next ten years the number of middle-income jobs traditionally held by women, such as teacher, librarian, and counselor, will decrease. In contrast, there will be rapid increases in two important areas—low-paying service jobs, such as dental and medical assistants, and higher-paying jobs in administration, engineering, and the computer field.[17] If the trend toward increasing female poverty is to be arrested, women will have to take their share of these higher-paying jobs. But will they be prepared? Will they obtain the mathematical training necessary to become computer designers and engineers to join the well-paid middle class of the Silicon Valley?

THE BIOLOGICAL CONNECTION

In answer to these questions, once again the question of biology looms large. In elementary school boys and girls do equally well in math, but in high school girls take fewer math courses and often do less well than do boys on standardized math tests. One school of thought argues that this behavior is a result of a complex series of social factors that impel girls to avoid the study of math. Another suggests that there really are no sex-related differences to begin with, and a third proclaims the likelihood of innate sex differences in mathematical ability.[18] This last point of view implies, of course, that under any conditions males will do better in math, making it likely that more males than females will become scientists, engineers, and computer and technical experts. The road to these sorts of high-paying jobs—in fields that will continue to expand in the 1980s—will remain partially blocked to women because of their supposed natural mathematical disability. Here, too, belief in a biological explanation for a social phenomenon suggests that efforts to change the existing situation are futile. Improve mathematics and science education? By all means, yes. But develop special programs to encourage girls to study math and science? Why?—it would be throwing good money after bad.

Should women vote and go to college? Are they governed by uncontrollable monthly rages? Can they compete in the job market? At the heart of all of these debates is an old question about human behavior. In the past people framed it in terms of nature versus nurture. Today it most frequently reaches us in phrases such as "genetic basis" or "genetic deep structure." The idea that our genes, factors inherited from our parents, somehow determine who we are and what we can become is now so widespread that even advertisers use it to hype their products. Consider the Calvin Klein ad[19] that starts with the headline "Thanks for the Genes, Dad!!" Since one good turn deserves another, the handsome, sexy children in this ad have decided to give Calvin Klein jeans to their father for Father's Day. (Get it? Jeans in return for Genes?)

Genetic punning aside, the question of nature versus nurture—or, as it is sometimes more ponderously phrased, biological versus social determinism—remains a hotly debated topic. Although many of the arguments explored in subsequent chapters of this book are posed in this either/or fashion, some scientists and social theorists (myself included) no longer believe in the scientific validity of this framework. Such thinkers reject the search for unique "root causes," arguing instead for a more complex analysis in which an individual's capacities emerge from a web of interactions between the biological being and the social environment. Within this web, connecting threads move in both directions. Biology may in some manner condition behavior, but behavior in turn can alter one's physiology. Furthermore, any particular behavior can have many different causes. This new vision challenges the hunt for fundamental biological causes at its very heart, stating unequivocally that the search itself is based on a false understanding of biology. The question, "What fraction of our behavior is biologically based,"[20] is impossible— even in theory—to answer, and unanswerable questions drop out of the realm of science altogether, entering instead that of philosophy and morality.

Science, according to definition, is knowledge based on truth, which appears as fact obtained by systematic study and precise observation. To be scientific is to be unsentimental, rational, straight-thinking, correct, rigorous, exact. Yet in both the nineteenth and twentieth centuries scientists have made strong statements about the social and political roles of women, claiming all the while to speak the scientific truth. Feminists, too, have used scientific arguments to bolster their cause.[21] Furthermore, research about sex differences frequently contains gross procedural errors. In a 1981 article one well-known psychologist cited "ten ubiquitous methodological problems" that plague such work.[22] The list contains striking errors in logic—such as experiments done only on males from which the investigators draw conclusions about females, and the use of limited (usually white, middle-class) experimental populations from which a scientist draws conclusions about *all* males or females. Perhaps the most widespread methodological problem is pinning the results of a study on gender when differences could be explained by other variables. Many researchers note, for example, that boys do better than do girls on college entrance tests in mathematics. For years scientists concluded from such results that boys are better at math than are girls. Recently, however, several investigators have pointed out that girls take fewer math courses in high school; thus college entrance exams pit boys with more training in math against girls with less training. Sex and course taking are confounded, and the conclusion that boys are inherently better at math remains without clear-cut support.

What is the untrained onlooker to make of all this? Are these examples of "science corrupted," as one historian has called the misrepresentation of women in scientific studies,[23] or do such cases provide evidence for a rather different view of science—one in which the scientists themselves emerge as cultural products, their activities structured, often unconsciously, by the great social issues of the day? During the past fifteen years scholars in women's studies have looked hard at virtually every field of intellectual inquiry, all the while feeling more and more like the child in the story about the emperor's new clothes. Examining the same material that for years great intellects had deemed solid, whole, flawless, they have found themselves asking, naïvely at first, but then with greater factual and theoretical sophistication, "But where are the women?" and, "If you take women into account, doesn't that change the whole conclusion?" Scientific inquiry, particularly as it pertains to sex and gender, has been no exception.

If science as an overall endeavor is completely objective and functions independently of the prevailing social winds, then scientists who commit gross errors of method and interpretation are simply bad at their jobs. The problem with this view is that flaws in research design often show up in the work of intelligent, serious men and women who have been trained at the best institutions in the country. By all conventional measures—publication record, employment in universities, invitations to scholarly conferences— they are good scientists, highly regarded by their peers. Here, then, we face an apparent paradox. Some of the most recognized scientists in their fields have built a reputation on what others, myself included, now claim to be bad work. One could resolve the paradox simply by denouncing the entire scientific enterprise as intellectually corrupt, but I find this an unacceptable position. I believe that the majority of scientists not only are highly capable but that they try in good faith to design careful, thoughtful experiments. Why, then, do they seem to fail so regularly when it comes to research on sex differences?

The answer may be found if, rather than simply dismissing these researchers as bad at their trade, we think about what they do as "conventional science." In analyzing male/female differences these scientists peer through the prism of everyday culture, using the colors so separated to highlight their questions, design their experiments, and interpret their results. More often than not their hidden agendas, non-conscious and thus unarticulated, bear strong resemblances to broader social agendas. Historians of science have become increasingly aware that even in the most "objective" of fields—chemistry and physics—a scientist may fail to see something that is right under his or her nose because currently accepted theory cannot account for the observation.[24] Although no one can be entirely successful, all serious scientists strive to eliminate such blind spots. The prospects for success diminish enormously, however, when the area of research touches one very personally. And what could be more personally significant than our sense of ourselves as male or female? In the study of gender (like sexuality and race) it is inherently impossible for any individual to do unbiased research.

What, then, is to be done? We could call for a ban on all research into sex differences. But that would leave questions of genuine social and scientific interest unanswered. We could claim an agnostic position—that all research is good for its own sake—but no one really believes that. Scientists make judgments all the time about the importance of particular lines of research, and those deemed frivolous or otherwise insignificant fail to receive funding. We ought, therefore, neither to impose research bans nor to claim agnosticism. Instead, we ought to expect that individual researchers will articulate—both to themselves and publicly—exactly where they stand, what they think, and, most importantly, what they *feel* deep down in their guts about the complex of personal and social issues that relate to their area of research. Then let the reader beware. The reader can look at the data, think about the logic of the argument, figure out how the starting questions were framed, and consider alternate interpretations of the data. By definition, one cannot see one's own blind spots, therefore one must acknowledge the probability of their presence and provide others with enough information to identify and illuminate them. In a sense, what I do throughout this book is take a flashlight and shine it in the unlit corners of other people's research.

Since I exhort other scientists to spell out their beliefs, to step out from behind the mask of objectivity, it behooves me to do the same. I have been trained to do laboratory research in a field technically designated as developmental genetics. The area represents a cross between embryology—the study of the development from single egg cells into complex, many-celled organisms—and genetics, the study of the mechanisms of inheritance. As a scientist I am inevitably a materialist. To my mind Western science provides a particular (but not the only) description of a material reality that in many respects makes sense to me. Despite my general acceptance of Western scientific thought, I find the available concepts and tools of modern science inadequate to describe certain kinds of reality, especially those that are multiply determined—that is, those for which a particular end may be reached by a number of alternate pathways. Complex social behavior is one example.

As a university faculty member I spend a considerable amount of time in the classroom sharing with students some of the things I know and, more importantly, helping them gain learning skills which they will continue to use long after they leave college. When I first began to teach I concentrated on helping my students learn about embryology. In recent years, however, I have begun to teach two new subjects; the biology of gender, in which we explore the expression of gender and its evolution throughout the animal and plant kingdoms, and a topical course that examines a variety of social issues in biology, including the controversy over race and intelligence and the sociobiology debate. Such courses clearly relate to another important aspect of my professional life—thinking and writing about the biology of gender in humans, about the sociology of science, and about feminism and science. These interests in turn stem from my life as a political activist. Since the late 1950s I have participated in the civil rights, antiwar (Vietnam), and women's liberation movements, all of which seemed to me to be a means toward reaching the goal of a world in which men and women of all races live in substantial social, economic, and political equality. My belief in such a future is of long standing and is deeply held.

The reader is by now in a position to ask me a tough question. I have mentioned scientists who fail to maintain their objectivity, suggesting that blind spots are an inherent aspect of science and are most frequent and dangerous when one studies socially urgent topics. In writing this book, am I guilty from a feminist standpoint of just what I accuse others of doing from a nonfeminist one? My answer, of course, must be no. In this book I examine mainstream scientific investigations of gender by looking closely at them through the eyes of a scientist who is also a feminist. Because of my different angle of vision, I see things about the research methods and interpretations that many others have missed. Once pointed out, much of what I have to say will seem acceptable, even to those whose research I criticize; but some of what I write will be controversial. In the end, the resolution of such controversy often depends upon one's standard of proof, a standard dictated in turn by political beliefs. I impose the highest standards of proof, for example, on claims about biological inequality, my high standards stemming directly from my philosophical and political beliefs in equality. On the other hand, given the same claims, a scientist happier with present-day social arrangements would no doubt be satisfied with weaker proof. How much and how strong the proof one demands before accepting a conclusion is a matter of judgment, a judgment that is embedded in the fabric of one's individual belief system.

In the pages that follow we will look closely at many scientific claims about men and women. We will start with the assertion that male and female brains differ physically, with the result that the members of each sex end up with different abilities for verbalizing and doing mathematical work. Since at the heart of this and other arguments is the idea that genes cause behavioral differences between the sexes, we will ask just what is meant by the idea of genetically caused behavior. At the same time we will also see what is known about the embryological development of gender differences. Women's hormonal ups and downs, some would say, make them emotionally unstable, while men's hormones make them the more aggressive sex. But is there scientific evidence to support such ideas? Finally, we will look at a body of thought that, within the framework of knowledge about human evolution, tries to find explanations for present-day male/female interrelationships. For each of these topics we will not only discuss the views of some scientists and physicians, we will ask just how well the scientific literature backs up a particular viewpoint.

This book is a scientific statement *and* a political statement. It could not be otherwise. Where I differ from some of those I take to task is in not denying my politics. Scientists who do deny their politics—who claim to be objective and unemotional about gender while living in a world where even boats and automobiles are identified by sex—are fooling both themselves and the public at large. Through their public pronouncements, through their articles in the *New York Times*, through their claims to hold an apolitical, scientific truth, through their efforts to construct biological connections where there are none, such scientists divert attention from some of the most pressing political issues of our time.

2

A QUESTION OF GENIUS:
ARE MEN REALLY
SMARTER THAN WOMEN?

There is perhaps no field aspiring to be scientific where flagrant personal bias, logic martyred in the cause of supporting a prejudice, unfounded assertions and even sentimental rot and drivel have run riot to such an extent as here.
—HELEN THOMPSON WOOLLEY
Psychologist, 1910

It would be difficult to find a research area more characterized by shoddy work, overgeneralization, hasty conclusions, and unsupported speculations.
—JULIA SHERMAN
Psychologist, 1977

JOBS AND EDUCATION—that's what it's really all about. At the crux of the question "Who's smarter, men or women?" lie decisions about how to teach reading and mathematics, about whether boys and girls should attend separate schools, about job and career choices, and, as always, about money—how much employers will have to pay to whom and what salaries employees, both male and female, can command. These issues have formed an unbroken bridge spanning the length of a century. Across that passageway, year in and year out, have trucked thousands upon thousands of pages written to clarify our understanding of the intellectual abilities of men and women. Hundreds of this nation's top educators, biologists, and psychologists have done thousands of studies offering us proofs, counterproofs, confirmations, and refutations. Yet the battle rages with as much heat and as little light as ever.

Today's claims are quite specific. The science feature page of the *Boston Globe* had the following headline in an article on education:

IS MATH ABILITY AFFECTED BY HORMONES? Far more boys than girls get top scores in math test.[1]

In the same vein a mathematics teacher in a Warwick, Rhode Island, high school writes:

As a mathematics educator with over 25 years in dealing with female pupils and female mathematics teachers, I do have direct evidence . . . mathematics is the water in which all intellectual creativity must mix to survive. Females, by their very nature, are oleaginous creatures in this regard. Or . . . as the song says: "Girls just wanna have fun."[2]

Theories abound that there are more male than female geniuses and that boys wind up ahead of girls in the classroom and hence in the job market. Why? Because, some would hold, hormonal differences between the sexes cause differences in brain structure and function. These in turn lead to differences in cognitive ability. Boys supposedly develop greater visual-spatial acumen; girls develop better verbal and communication skills. Although many researchers take such differences for granted, my own reading of the scientific literature leaves me in grave doubt about their existence. If sex differences in cognition exist at all they are quite small, and the question of their possible origins remains unanswered. Nevertheless, the claim of difference has been and continues to be used to avoid facing up to very real problems in our educational system and has provided a rationale for discrimination against women in the workplace. The issue of cognitive differences between the sexes is not new. Scientists and educators used versions of this particular scientific tale even before the turn of the century.

In 1903 James McKeen Cattell, a professor at Columbia University and editor of *Science*, the official journal of the American Association for the Advancement of Science, noted that among his list of one thousand persons of eminence throughout the ages, only thirty-two were women. Although Cattell expressed some surprise at the dearth of eminent females, he felt that it fit with the fact that in his *American Men of Science* only a tiny number of women

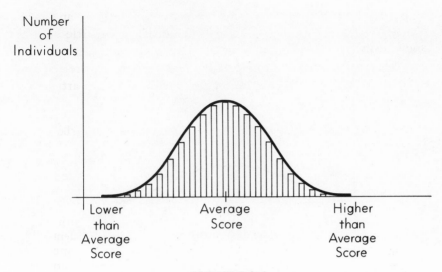

FIGURE 2.1
Genesis of the Bell-shaped Curve

appeared among the top thousand scientists. From his standpoint "there [did] not appear to be any social prejudice against women engaging in scientific work," hence he found it "difficult to avoid the conclusion that there is an innate sexual disqualification."[3] Another Columbia professor, Edward L. Thorndike, an influential educational psychologist and a pioneer in the use of statistics in educational research, also commented on the lack of intellectually gifted women. As an advocate of educational efficiency, he saw little sense in squandering social resources by trying to train so many women to join the intellectual elite. An exceptional female could become an administrator, politician, or scientist, but the vast majority were better off learning to become nurses and teachers where, as he put it, "the average level is essential."[4]

Thorndike and Cattell both thought that biological differences between the sexes explained the rarity of extremely intelligent women. Men, it seemed, were by nature more variable and this variability created more male geniuses. Since the line of reasoning may at first seem tortured, a word of explanation is in order. Researchers give tests to groups of individuals. If one displays the number of people with a particular test score on a graph, as shown in figure 2.1, the distribution of performances usually approximates a bell-shaped curve. The highest part of the curve, showing the scores most frequently attained, represents the average performance.

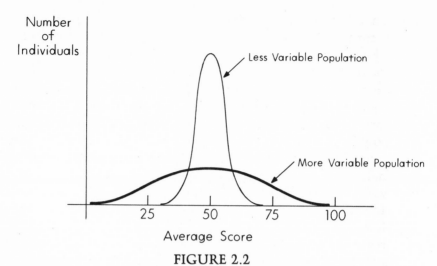

FIGURE 2.2

Bell-shaped Curves of Populations with the Same Average Trait
but with Different Degrees of Variability

Individuals whose scores fall to the right have performed above average while those whose scores fall to the left were below average.

There is, however, more than one way to reach an average. On a test in which the highest possible score is 100, for example, the average might be 50. If the average resulted from the fact that everyone scored very close to 50, the bell-shaped curve would be very tall and narrow. If, on the other hand, the average score of 50 resulted from a population of individuals, some of whom scored in the 90s and some of whom scored in the teens, the shape of the bell would be low and squat (see figure 2.2). In the former example, where all of the individual scores hover right around the group average, the *standard deviation from the mean* is small, while in the latter case it is quite large. A population with a large standard deviation is, quite obviously, highly variable, making it harder to predict the performance of any one individual in the group.

What does all this have to do with an excess of male geniuses? Thorndike and others agreed that men and women had the same *average* level intelligence. But men were more variable; thus, their intelligence curve looked more like the short, squat one drawn in figure 2.2, while the women's looked more like the tall, narrow one. (I've exaggerated the effect to illustrate the point more clearly.) What counted for the men was the above-average tail on the bell curve, containing as it must individuals who surpassed the abilities of even the most gifted women. (The variability hypothesis also allows for the presence of a greater number of subnormal males, a fact acknowledged by both Thorndike and Cattell.)

The theme of variability is an old one. Before Darwin published his theory of evolution, Western scientists considered variability a liability to the species. They also thought that *women* constituted the more variable sex. Darwin, however, won credence for the ideas that populations with greater variability among individuals had a better chance of withstanding the evolutionary test of time, and that males were more likely to vary than were females. Thus, when high variability was considered to be a biological drawback, it was attributed to the female of the species; in its post-Darwinian status as a biological benefit, it became a male property and males remained the progressive element, the active experimenters of their race. In return for relinquishing their variability, women received the mantle of conservation, becoming the passive vessels of racial purity.[5]

A number of psychologists working in the first quarter of the twentieth century published competent scientific studies disputing the claim that men were more variable than women.[6] But the ideas of Thorndike and others that women should be educated for professions such as nursing, social work, and teaching were backed by powerful social forces. Cattell, for example, wrote at a time when women had begun to outnumber men as students in many of the large state universities—California, Iowa, Minnesota, and Texas among them.[7] The "problem of feminization" concerned educators deeply. While administrators at the University of Chicago contained the large growth in women students by placing them in a separate college within the school, other institutions responded by urging women to enter special all-female fields. Home economics, for one, provided a new place for the increased number of women chemists.[8] The structure of the work force had also changed markedly. With job segregation a fact of life for women,[9] Thorndike and others encouraged the massing of women into certain (low-paying) occupations by urging the utility of separate vocational education for males and females. Federal aid for industrial arts programs for boys and home economics courses for girls supported this process. According to one analysis, "Hospital and school administrators welcomed these programs as a solution to their growing need for competent but inexpensive workers. Businessmen supported the growing number of secretarial and commercial courses for women for similar reasons."[10] The biological views of Cattell and Thorndike were so congenial to the economic and political establishment of the period that rational, scientific challenges to their work were studiously ignored.[11]

The debate over variability went on into the 1930s, when it finally seemed to have been laid to rest by Lewis Terman, an expert on mental testing.[12] But, like the phoenix arising fresh and beautiful from the ashes of its own cremation, the theory of variability has appeared once more on the modern scene. Curiously, its rebirth brings out few new facts, presenting only a somewhat modernized formulation of the same old idea. In 1972 the *American Journal of Mental Deficiency* published an article by Dr. Robert Lehrke entitled "A Theory of X-linkage of Major Intellectual Traits."[13] The editors were sensitive to the fact that the article would provoke controversy and took the somewhat unusual step of inviting three well-known psychologists to write critiques, which followed the original article along with a round of response from Lehrke.[14] Lehrke noted that there were more institutionalized mentally retarded males than females, an observation made by many but poorly understood.[15] Is it possible that parents keep retarded girls at home more often? Are boys more susceptible to environmental shock? Or, does the X-linkage of certain metabolic diseases* make boys more likely to be institutionalized? Lehrke's hypothesis holds that a number of genes relating to intellectual ability reside on the X-chromosome and that, because of the peculiarities of chromosomal inheritance, X-linkage means that males will exhibit greater variability in intelligence. Although he begins with the supposed excess of mental defectives, Lehrke does not shrink from the implication that there would also be more genius-level males. As he rather succinctly wrote: "It is highly probable that basic genetic factors rather than male chauvinism account for at least some of the difference in the numbers of males and females occupying positions requiring the highest levels of intellectual ability."[16]

To understand some of the details of Lehrke's argument it is worthwhile to review the idea of X-linkage. Males and females differ genetically. In addition to twenty-two pairs of chromosomes called autosomes, females have two X chromosomes. Males, on the other hand, supplement their twenty-two autosomes with one X and one Y chromosome. Because X and Y chromosomes are associated with the development of gender, they are sometimes referred to as

---

* Hemophilia, for example, is X-linked and therefore affects boys more frequently than girls.

the sex chromosomes. Hemophilia, a particularly famous X-linked disease, illustrates the process of X-linked inheritance. The hemophilia gene, which resides on the X chromosome, exists in two states—normal and mutant. The normal gene codes for a factor that helps blood to clot, while the mutant gene cannot aid in the production of the clotting factor. Since males carry only one X chromosome, and since the Y chromosome cannot counteract the effect of genes on the X chromosome, a male will suffer from hemophilia if he carries an X chromosome with the mutant state of the gene. A female must carry two abnormal X chromosomes in order to be a bleeder, because she will be protected as long as one X chromosome carries the normal gene.

Children, however, can inherit the mother's abnormal X chromosome. Since sons derive their X chromosomes from their mothers, a mother carrying the hemophilia factor on one of her X chromosomes stands a 50 percent chance of having a hemophiliac son. On the other hand, since daughters receive one X chromosome from the mother and the other from the father, a stricken girl must have a hemophiliac father in addition to a carrier mother. In other words, if hemophilia runs in the family, sons will express the trait more frequently than will daughters.

Lehrke hypothesizes that, unlike the clotting factor gene that exists in one of two possible states, an X-linked gene for intelligence might exist in as many as six graded states—called *alleles* in the terminology of geneticists—running from lower to higher intelligence. A female would always carry two of these (one on each X chromosome), and the one evoking greater intelligence might then compromise with the one for lesser intelligence. Males, on the other hand, would only carry one allele at a time. If that one allele coded for a low state of intelligence, then the male would express that trait, while if the allele were one for the highest state, the individual would be extremely intelligent. According to Lehrke, then, one would find equal levels of retardation or genius among males and females. However, because expression of extremes of intelligence in females would require *two* chromosomes with the same very low or very high state of brightness, while expression in males would require the presence of only one, a larger number of males than of females would be found who were either extremely dull or incredibly brilliant. Hence, the greater male variability in intelligence.[17]

The most fundamental assumption in Lehrke's hypothesis is that intelligence is an inherited trait coded for by some finite number of factors called "intelligence genes." This claim has evoked great controversy, and many well-known biologists have argued convincingly that (1) it is impossible to define intelligence, and (2) we have no means at our disposal to measure its genetic component separately from its environmental determinants.[18] Lehrke bolsters his argument by citing the work of Arthur Jensen, who figures heavily in a long-standing debate over whether blacks are less intelligent than whites. Jensen and others believe that whites are smarter and that educational enrichment programs for underprivileged children are a waste of government money. From his comments about the lower intelligence of slum dwellers, one would suspect that Lehrke agrees with this concept.[19] Lehrke also claims that the existence of several X-linked traits that cause mental retardation proves the inherited nature of intelligence. This argument includes the hidden, circular assumption that mental deficiency results from genes specific to the development of intelligence. My point can be illustrated by looking at one often cited example, the disease called phenylketonuria (PKU).

In the very recent primary literature the simple autosomal inheritance of PKU has been called into question,[20] but virtually all genetic and medical textbooks use this disease as an example of the straightforward inheritance of a gene that "causes" mental retardation. Children born with PKU lack an enzyme called phenyl alanine hydroxylase, which converts the amino acid phenyl alanine—one of the building blocks of large protein molecules—to another amino acid, called tyrosine. Because their cells cannot make this conversion, PKU patients accumulate toxic levels of phenyl alanine—from forty to fifty times the normal amount—in the blood and brain. Since the brain continues to develop actively even after birth, its cells may be particularly sensitive to this poison. Indeed, children with PKU fed on a diet lacking in phenyl alanine develop fairly normally.

The question is whether the existence of the inherited disease phenylketonuria (or similar diseases of metabolism) provides evidence that genes govern intelligence. That normal intelligence requires normal brain development is obvious, but the existence of PKU says nothing about the presence of genes for intelligence or learning. It merely says that when the entire brain is poisoned during a critical period of development, the effects can be disastrous. From the point of view of explaining the relationship between genes and intelligence, this is no more informative than asserting, after smashing someone's head with a sledgehammer, that violence done to a person's skull causes subsequent mental dullness. The same point can be made for all of the gene and chromosome defects that severely affect normal human development. They give us a glimpse of what can go wrong, but they tell us absolutely nothing—at least in terms of intelligence—about how things work right.

Arguments against the idea of intelligence genes seem sufficient to warrant dismissal of Lehrke's hypothesis.[21] But he both resurrects old data and cites newer information purporting to show once again that males perform more variably on intelligence tests than do females, and those citations merit consideration. Investigating variability in IQ turns out to be a rather formidable task. In one recent study researchers looked for scientifically gifted children by holding math and science contests. They found a greater number of precocious boys than girls, and their top winners were all male. They noted, however, that among the precocious students the boys owned more books and equipment related to math and science, while some of the girls' parents were so uninterested in their daughters' precocity that they didn't even plan to send them to college.[22] The existence of such social differences between the boys and the girls makes the results difficult to interpret. Furthermore, a "talent search" approach looks only at a select group of students who either volunteered for the contests or were recruited by teachers or parents. Although Lehrke doesn't cite this study, he does cite an older one[23] which is subject to the same sorts of uncertainty.

The only way to get some sense of the variability of the population as a whole is to do large-scale, nonselective studies. These are expensive and difficult to design—there is only one well-done research project of this kind in the literature. Lehrke cites this project, a survey of Scottish schoolchildren, to support his view that males vary more than do females. Since the sample was very large, and since sample size is one of the components statisticians use to decide whether a particular difference is significant rather than just random, the small differences in standard deviation found among these Scottish boys and girls turned out to be statistically significant. As one of the respondents to Lehrke's 1972 paper points out, however, the male variability resulted mostly from an excess of males with very *low* scores—a result, perhaps, of physical handicaps that might have interfered with their performance on the timed tests.[24]

Lehrke's response to his critics is maddening. He concedes that "each one of the arguments for X-linkage of major intellectual traits can be interpreted to produce different emphases," but thinks that his emphasis merits attention because it is a simpler explanation.[25] In addition to this weak attempt at scientific rebuttal, and despite the fact that some of his critics are male,[26] Lehrke also directly points to what he thinks is the real source of his trouble:

> Determinants of which viewpoint a person accepts are undoubtedly highly complex, but a single, very simple one is obvious. In the small sample cited, all those accepting the hypothesis of greater male variability have been males, all those rejecting it, females.[27]

In contrast to his assessment of his female critics, however, Lehrke believes himself to be a dispassionate observer:

> I do not feel that I must apologize for the fact that certain implications of the theory may seem . . . to be derogatory to women. Like Topsy, the theory "just growed," its nature being determined by the data. I could not, with scientific objectivity, have changed the final result.[28]

Here, then, we have the elements of a response that will show up again and again in debates touched upon throughout this book. In each case, the proponents of biological explanations of behavior label their attackers as biased, members of some special interest group (women, feminists, Marxists), while choosing for themselves the role of the objective, dispassionate scientist.

Before judging Lehrke's detachment, though, the reader ought to know a little something about the company he keeps. His last article, "Sex Linkage: A Biological Basis for Greater Variability in Intelligence," was published in 1978 in a book entitled *Human Variation: The Biopsychology of Age, Race, and Sex.*[29] The book is dedicated to the memory of Sir Francis Galton, founder of the eugenics movement, while its headquote comes from none other than E. L. Thorndike. Just as interesting, the volume in question is edited by Dr. R. Travis Osborn, a leader in the new eugenics movement,[30] who has received, over the years, financial support from a "philanthropic" organization called the Pioneer Fund, which promotes theories of black inferiority and has supported the work of Drs. William Shockley and Arthur Jensen. (Past members and directors include Senator James O. Eastland, the segregationist senator from Mississippi, and Representative Francis E. Walter, who chaired the House Committee on Un-American Activities during the anticommunist campaigns of the 1940s and 1950s.[31]) Are Lehrke, Osborne, and Jensen (who also has an article in the book) strange bedfellows or, as I suspect, appropriate company—each being a scientist who disclaims responsibility for the social implications of his "objective" facts?

Cattell and Thorndike formed part of the mainstream of educational psychology which to this day carries along such adherents as Lehrke. There were others in the mainstream who rejected the variability hypothesis but argued instead that innate differences between males and females are important when considering what jobs to train for and how to teach—even at the elementary level— such subjects as reading and writing. Among the most widely quoted compilations of data on sex differences is one published in 1968 by Garai and Scheinfeld. In the introduction to their book-length literature review they explicitly state that their purpose is "to make the participation of women in the labor force as efficient as their potential permits." To summarize Garai and Scheinfeld's findings in their own words:

> Females, on the average, surpass males in verbal fluency, correct language usage, spelling, manual dexterity, clerical skills, and rote memory. Males, on the average, are superior to females in verbal comprehension and verbal reasoning, mathematical reasoning, spatial perception, speed and accuracy of reaction to visual and auditory stimulation, mechanical aptitude, and problem-solving ability. *These sex differences foreshadow the different occupational goals of men and women.*[32] [Emphasis added]

In a conclusion echoed in more recent writings by other psychologists, Garai and Scheinfeld infer that women's work preferences lie in the fine arts, literature, social services, secretarial jobs, and assembly-line work because these areas suit their particular aptitudes. Men, in contrast, seem drawn by their special skills to the sciences, mathematics, engineering, mechanics, and construction.

Garai and Scheinfeld call for certain educational reforms to accommodate their findings. They believe that boys are handicapped by coeducational classes because they mature more slowly, while girls are distracted, especially in more difficult subjects, by their need for approval and interaction with others. Their solution would be a return to single-sex classes, at least in high school. Garai and Scheinfeld also suggest that there are separate feminine and masculine ways of learning subjects such as mathematics and reading, and that teaching methods for these subjects ought to be reevaluated. This thought, too, remains current. In a research paper appearing in

Science magazine in 1976,[33] psychologist Sandra Witelson concludes that boys' and girls' brains have different physical organizations and that current methods of teaching reading (which stress phonetics rather than visual memory) may favor girls while handicapping boys. In an interview she, too, said that "separate groups or classes for the sexes would be beneficial for teaching reading."[34]

In this day of increasing coeducation, the thought of resegregating classrooms by sex carries a certain irony. At the college level there *is* evidence that coeducation as currently practiced may harm *female* students.[35] But is the solution to return to a separate but unequal form of education,[36] or to identify and remedy whatever it is about coeducation that functions to discourage female students? Of course if one believes in innate sex differences, then the latter makes no sense.

With echoes of James McKeen Cattell in our ears, we find ourselves once again in a period in which females outnumber males on the college campuses. In the current political climate the enrollment changes have led not to a move to cordon off the females as in Cattell's day, but instead to a call from students for more female faculty and better role models. Garai and Scheinfeld, however, call for the "defeminization of the elementary classroom."[37] There are, they feel, too many women teachers whose emphasis on conformity and good behavior stifles the creative expression of little boys; girls, too, need more male teachers, especially if they are to be encouraged (at least the more talented ones) to study science or to improve their creative abilities. Garai and Scheinfeld claim that "almost exclusive staffing of libraries with women and of schools with women teachers create[s] a climate which confronts the boy with hostility and lack of understanding," curiously echoing a diatribe written by Cattell in 1909 in which he, too, deplored the dominance of the female "school principal, narrow and arbitrary, and the spinster, devitalized and unsexed" over the school lives of little boys and girls.[38] Thus, while feminists call for more female role models, some psychologists call for a return to the male-dominated classroom. Is there truly no scientific evidence to tell us who is right?

TABLE 2.1
Summary of Maccoby and Jacklin's Findings on Sex Differences

Unfounded Beliefs About Sex Differences	*Open Questions of Difference*	*Fairly Well Established Sex Differences*
Girls are more social than boys	Tactile sensitivity	Girls have greater verbal ability
Girls are more suggestible than boys	Fear, timidity, and anxiety	Boys excel in visual-spatial ability
Girls have lower self-esteem than boys	Activity level	Boys excel in mathematical ability
Girls are better at rote learning and simple repetitive tasks; boys are better at higher level cognitive processing	Competitiveness Dominance Compliance Nurturance and "maternal" behavior	Boys are more aggressive
Boys are more analytic than girls		
Girls are more affected by heredity; boys are more affected by environment		
Girls lack achievement motivation		
Girls are more inclined toward the auditory; boys are more inclined toward the visual		

SOURCE: Eleanor Maccoby and Carol Nagy Jacklin, *The Psychology of Sex Differences* (Stanford, Calif.: Stanford University Press, 1974).

Male Skills/Female Skills: The Elusive Difference

The best starting point for discussing the difference between male and female skills is a book published in 1974 by two psychologists, Eleanor Maccoby and Carol Nagy Jacklin.[39] They summarize and critically evaluate a large body of work on the psychology of sex differences, concluding that at least eight different claims for sex differences (see left-hand column in table 2.1) were *disproved* by the results of then available scientific studies and that the findings

25

about seven other alleged differences (see middle column) were either too skimpy or too ambiguous to warrant any conclusions at all, but that sex differences in four areas—verbal ability, visual-spatial ability, mathematical ability, and aggressive behavior—were "fairly well established" (see right-hand column). We turn our attention for the remainder of this chapter to the first three of these differences: verbal, visual-spatial, and mathematical abilities. The fourth, aggressiveness, we will consider in chapter 5.

Verbal Ability

Many people believe that little girls begin to talk sooner than do little boys and that their greater speaking abilities make girls better able to cope with the word-centered system of primary education. Maccoby and Jacklin cite one summary of studies done before 1950 that points to a trend of earlier vocalization in girls. The gender differences, however, are small and often statistically insignificant, and, in fact, many of the studies show no sex-related differences at all. In their review of the literature subsequent to 1950, Maccoby and Jacklin remain skeptical about the existence of sex differences in vocalization for very young children. Although a small body of more recent work suggests that there probably *is* something to the idea that girls talk sooner than boys,[40] my own assessment is that the differences, if any, are so small relative to the variation among members of the same sex that it is almost impossible to demonstrate them in any consistent or statistically acceptable fashion.

The studies on early vocalization raise several interrelated issues in basic statistics that must be understood in order to delve further into the controversies surrounding verbal and spatial abilities. Among these issues are statistical significance and its relationship to sample size and the size of differences *between the sexes* compared with the size of differences *between any two individuals* of the same sex. This latter issue, of which psychologists in the field of cognitive differences have become increasingly aware, places the importance of sex differences in a whole new light.

One widely accepted scientific procedure for comparing averages obtained from individual measurements of members of a population is to apply a statistical test to the information gathered. An average difference between two groups could occur by chance and therefore would reflect no real distinction between the two test populations.

A QUESTION OF GENIUS

A random difference is particularly likely if the individual trait under study in one or both of the groups varies a lot (that is, has a large standard deviation). Scientists have devised several methods for examining experimental information to find out if average differences are real rather than chance. Most such statistical tests look at two things—the variability of the populations under comparison and the size of the test sample. If a test sample is very small, very variable, or both, the possibility that found differences are due to chance is great.

Suppose, for example, I suspect that more males than females have blue eyes. In order to test my idea, I look at three groups of ten students (five men and five women) borrowed from three different classrooms. In the first group it turns out that two-thirds of the men but only one-third of the women have blue eyes, in the second that two-thirds of the women but only one-third of men have blue eyes, while in the third classroom all five of the men have blue eyes, but none of the women do. Taking the average of my three samples, I see that, overall, 66 percent of the men have blue eyes compared to only 33 percent of the women.

In standard scientific convention one tries to discover the probability that a particular result can occur by accident. Because my sample in the preceding example was small and variable, this probability was 65.6 percent (calculated using a special statistical test that takes into account variance and sample size). Scientists use an agreed-upon albeit arbitrary limit, whereby a hypothesis is rejected if the probability of a found difference occurring by chance exceeds 5 percent. Thus I must reject the hypothesis that more boys than girls have blue eyes, as it is based on a poor data sample. If the probability of a difference existing by chance is 5 percent or less, then one accepts the hypothesis and calls the results *statistically significant*.

Statistical significance, however, can mislead, because its calculation comes in part from the size of the measured sample. For example, in order to show that two groups differ in performance by four IQ points, one must use a sample size of about four hundred in each test group (that is, if the sexes were compared, four hundred boys and four hundred girls). Greater differences in IQ can be shown with smaller groups, while *extremely large* samples may reveal statistically significant results, according to the convention of 5 percent probability, even though they are intellectually meaningless. Thus, in a sample of 100,000 males and 100,000 females,

TABLE 2.2
*Sex-related Cognitive Differences: Verbal Reasoning in Subjects
over Age Sixteen*

Variable	Number of Results	Female Superior	Male Superior	No Difference
Oetzel (1966)				
Vocabulary	4	2	0	2
Verbal Problem Solving	1	1	0	0
General Verbal Skill	4	1	0	3
Abstract Reasoning	4	1	1	2
Maccoby and Jacklin (1974)				
Verbal Abilities	25	8	1	16
Droege (1967)	2	2	0	0
TOTAL	40	15	2	23

NOTE: Julia Sherman, *Sex-Related Cognitive Differences: An Essay on Theory and Evidence* (Springfield, Ill.: Charles C Thomas, 1978), 40. Courtesy of Charles C Thomas, Publisher.

an IQ difference of 0.02 points would be highly significant (probability of 0.1 percent). But it doesn't actually matter if one person has an IQ of 100 and another an IQ of 100.02, because the IQ test is not designed to measure such small differences.[41]

Using a somewhat unusual statistical manipulation, Garai and Scheinfeld concluded that girls were poorer at verbal reasoning than were boys. In order to reach that conclusion, they used the following approach. They knew that boys matured physically at a slower rate than did girls. In studies done on children of the same age, then, they believed the girls to be physically more mature and thus not *really* age-matched with the boys. They reasoned, therefore, that any of the studies that showed boys and girls to perform equally actually provided proof of male superiority![42] One way to get around the problem of different maturation ages is to look carefully at the studies done on people over the age of sixteen, a point in the life cycle at which the large majority of both boys and girls have gone through puberty. Dr. Julia Sherman has done just this. Her results, reproduced in table 2.2, show that in forty different studies of verbal reasoning done on subjects over the age of sixteen, females did better in fifteen and males in two, while in twenty-three there were no sex-related differences.

Two observations can be made from this information. First, when there *are* sex-related differences in verbal reasoning, females usually come out ahead. Second, in the majority of cases there are no differences at all. What, then, is the take-home message? Maccoby and Jacklin chose to emphasize the female superiority in the cases where there is some difference. They are, however, perfectly aware that the frequent inability to find any difference could be quite important. Given these data, choosing to believe in sex-related differences in verbal ability is a judgment call about which knowledgeable scientists can very legitimately differ.

More recently several researchers have related the difficulty of showing differences in verbal ability to the small size of any such differences. All the papers reviewed by Maccoby and Jacklin used what is called the *hypothesis-testing approach* to the study of sex differences. Using this approach, a researcher hypothesizes the existence, for instance, of a difference in verbal ability between boys and girls. Tests are given, average scores for boys and for girls are calculated, and the means, the standard deviations, and the number of subjects used to measure the statistical significance of any difference are presented. Maccoby and Jacklin simply tabulated how frequently a particular significant difference showed up in such studies.

Since the publication of their book, however, a new approach known as *meta-analysis* has been used by Jacklin and others to reevaluate their 1974 conclusions.[43] The new approach looks at the *size* of group differences, thereby allowing questions about such matters as verbal ability to be phrased in the following way: "If all you knew about a person was his or her score on a test for verbal ability, how accurately could you guess at his/her sex?" Meta-analysis is a highly sophisticated way of evaluating the meaning of several interrelated studies. It is simple in principle, albeit statistically complex. Instead of calculating separately the averages and standard deviations of males and females, one looks at the entire population (males and females together) and estimates the variability in the population as a whole using a statistic called the *variance*, which is related to the standard deviation.[44] Like the standard deviation, the variance tells one about the appearance of the bell-shaped curve that summarizes individual scores. In meta-analysis, one calculates how much of the variance found in the mixed population can be accounted for on the basis of gender, and how much is due to variation between members of the same sex and/or experimental error. We have already seen with the hypothesis-testing approach how one can obtain a meaningless but statistically significant difference by using a very large sample size. Meta-analysis provides a way of telling how large a given statistical difference is and thus how meaningful it is in reality.

Using meta-analysis, then, what becomes of Maccoby and Jacklin's "well-established sex difference" in verbal ability (see table 2.1)? It teeters on the brink of oblivion. Dr. Janet Hyde, for instance, calculated that gender differences accounted for only about 1 percent of the variance in verbal ability, pointing out that the tiny size of the difference could explain why so many of the studies cited by Maccoby and Jacklin show no difference at all.[45] Two other psychologists, Drs. Robert Plomin and Terry Foch, come to the same conclusion: "If all we know about a child is the child's sex, we know very little about the child's verbal ability."[46] Clearly, it makes little sense to base educational and counseling decisions that relate to verbal ability on simple observation of a child's sex, rather than on some actual analysis of his or her particular capacities.

Visual-Spatial Perception

"Males," one well-known psychologist has said, "are good at maps and mazes and math. . . . Females, by contrast, are sensitive to context."[47] Alliterative, yes, but is it true? Again, Maccoby and Jacklin provide the starting point. As with verbal ability, they conclude, there are no sex-related differences in visual-spatial abilities until adolescence. A summary of their findings from studies done on adolescents and adults appears in table 2.3. Spatial ability turns out to be somewhat elusive, but Maccoby and Jacklin have isolated two types: spatial/visual/nonanalytic and spatial/visual/analytic. Some scientists refer to this latter skill as *field articulation*.

The evidence for sex-related differences in visual-spatial ability seems a little more convincing than that for verbal differences, but the problem of "negative" data appears with both. More than half the time no sex differences show up in the visual/analytic studies, but when they do appear they always favor males. The most consistent differences materialize from the most widely used test, the rod and frame test. In this test the subject sits in a totally dark room in a chair facing a large (forty inches on a side), vertically held, luminescent frame. Bisecting the frame is a lighted rod. In one version the experimenter tilts the frame in various ways and the subject adjusts the rod to the vertical of the room, ignoring the immediate context of the tilted frame. In a different version, the

TABLE 2.3
Spatial Abilities of Adolescents and Adults

Skill	Number of Studies in which Males Performed Better	Number of Studies in which Females Performed Better	Number of Studies for which No Difference Was Found	Total
Visual/Nonanalytic[a]	8	0	2	10
Visual/Analytic				
Rod and Frame or Similar				
Test[b]	7	0	5	12
Embedded/Hidden				
Figures Test	3	0	6	9
Block Design Tests	2	0	2	4
Percentage for Visual/				
Nonanalytic	80%		20%	
Percentage for Visual/				
Analytic	48%		52%	

SOURCE: Eleanor Maccoby and Carol Nagy Jacklin, *The Psychology of Sex Differences* (Stanford, Calif.: Stanford University Press, 1974), tables 3.7 and 3.8.
[a] A variety of different tests involving mazes, angle matching, and 2- and 3-D visualization were used. The same test was rarely used twice.
[b] Body attitude test.

subject's chair is tilted, and again he or she must make the rod inside the frame perpendicular to the floor. As seen in table 2.3, Maccoby and Jacklin cite twelve studies using this test. Although women never performed better than men, in five of the twelve cases there were no sex-related differences.

Dr. H. A. Witkin, the psychologist who developed and popularized the rod and frame test, dubbed those who performed them well *field independent* and those who performed them poorly *field dependent*. Field-dependent people, Witkin and his collaborators held, were less able to ignore distracting background information in order to zero in on essentials. They suggested a relationship between general intelligence, analytical ability, conformity, passivity, and visual-spatial abilities. More recently, the fact that field-dependent and field-independent personalities just happen to correlate with male/female stereotypes has led a number of investigators to drop the use of the terms. It is now clear that these two tests, at best, record some aspect of visual skill, but have nothing to do with analytical ability. Witkin himself gave a tactile version of a test designed to measure field dependence to blind men and women and, except for one case favoring females, found no sex-related differences.[48]

Some potential for sex bias is built into the rod and frame test. Picture the following: a pitch dark room, a male experimenter, a female subject. What female would not feel just a little vulnerable in that situation? Although one would expect experimenter-subject interactions to be different for males and females in such a set-up, the studies cited by Maccoby and Jacklin apparently don't take into account this possibility. In one version of the test, the subject must ask the experimenter to adjust the rod by small increments to the position he or she believes to be vertical. A less assertive person might hesitate to insist to the nth degree that the experimenter continue the adjustments. Close might seem good enough. If it is true that females are less assertive than males, then this behavioral difference, rather than differences in visual-spatial acuity, could account for their performances in the rod and frame test. At least one experiment suggests the sex bias of the rod and frame test. When, in a similar test, the rod was replaced by a human figure and the task described as one of empathy, sex-related differences in performance disappeared.[49]

The rod and frame test is probably the most suspect of the measures used to assess male/female differences in spatial visualization, but psychologists use other tools as well to measure such skills. In the embedded figures test, the experimental subject must find a hidden word or design within a larger background that camouflages it. Another measure, the Wechsler Intelligence Scales, is used to assess IQ and comprises two tests, one measuring Verbal IQ and the other measuring Performance IQ. The latter is often taken as an indication of spatial ability, although some psychologists believe it to be inadequate for that purpose.[50] Other tests, some of them components of the standard IQ test, are used to probe the ability to visualize three-dimensional figures in the mind's eye. These include the block design test, the mental rotation test, angle-matching tasks, and maze performance. Psychologists have used all of these tests with rather similar results: many times no sex difference appears but when it does, and if the subjects are in their teens or older, males outperform females. The next question is, of course, by how much?

Maccoby and Jacklin point out that, as with differences in verbal skills, differences in spatial skills are quite small—accounting for no more than 5 percent of the variance. Expressed another way, if one looks at the variation (from lowest to highest performance) of spatial ability in a mixed population of males and females, 5 percent of it at most can be accounted for on the basis of sex. The other 95 percent of the variation is due to individual differences that have nothing to do with being male or female.[51]

Despite the small size of the difference, an advocate of the idea that there are naturally more male than female geniuses would have one strong point to make. If one looks at the entire bell-shaped curve, from worst to best, a small sex difference may be of no practical interest. Suppose, though, one looks only at the upper part of the curve, the portion representing those high-level performers one would expect to become math professors, engineers, and architects. Assume for a moment that in order to become a respected engineer one must have a spatial ability in at least the ninety-fifth percentile of the population. Dr. Hyde calculates that 7.35 percent of males will be above this cutoff in comparison with only 3.22 percent of the females. Put another way, currently available information suggests that the ratio of males with an unusually high level of spatial skills to that of females with the same high level of skills might be 2:1, a much larger difference than one picks up by looking at the entire population.[52] Hyde also points out, though, that in the United States only about 1 percent of all engineers are women. If one *did* believe that the only thing standing in the way of an engineering career for women was their immutable sex-related inferior spatial ability, one would still expect to find women in about one-third of all engineering jobs. In short, the differences between men and women in this respect remain too small to account for the tiny number of women who become professional mathematicians, architects, and engineers.

What Makes a Difference?

Sex differences in spatial visualization *do* sometimes exist, even if they don't amount to much. Thus there is an obligation to look into the causes of measurable difference. Because sex-related differences in verbal or spatial abilities appear most clearly at the time of puberty, some scientists conclude that the hormonal changes associated with physical maturation must affect male and female brain development differently. Others point to the social pressures to conform to appropriate role behavior experienced so intensely by adolescents. As seems so often to be the case, the same observation can support both a hypothesis of "natural," genetically based difference and one that invokes environmental influences. Couched in these terms, however, the clash of views has all the earmarks of a sterile, even boring, debate. Without trying to resolve competing hypotheses, let's simply look at the information we have at hand about the development of verbal and spatial abilities in little boys and little girls.

To begin with, there is ample evidence that visual-spatial abilities are at least in part learned skills. As an example, consider the fact that first-grade boys do somewhat better than do first-grade girls on embedded figures and blocks tests if neither has seen such tests before. Allowed a bit of practice, however, the girls improve enough to catch up, although the boys' scores do not change much. Researchers conclude from such studies that first-grade boys have already honed these skills so that additional practice does not lead to improved performance.[53] Why boys might be more practiced is anyone's guess, but since young boys and girls have quite different play experiences, one can at least construct a plausible hypothesis. Traditional male games such as model construction, block building, and playing catch might play a key role in developing visual-spatial skills, yet the relationship between play activities and the acquisition of spatial abilities has received scant attention from the research community.

Studies done on older children also reveal that three-dimensional visual skills can be learned. In one case a researcher assessed the performance of teenage students as they began a drafting course. The expected sex differences were found, but disappeared six weeks into the semester as the young women improved.[54] In another case teenagers showed a positive correlation between performance on tests of visual-spatial skill and the number of drafting and mechanical-drawing courses taken.[55] The sparse literature on the relationship between formal skill training (through certain types of course work) and informal (through certain types of play) suggests that girls often do not fulfill their skill potential, but that it would be relatively easy to help them do so. The hypothesis that certain kinds of play and school activities can improve girls' visual-spatial skills is eminently testable, but more research support is needed for scientists who are interested in carrying out such investigations.

It seems unlikely, however, that play and mechanical drawing are the only contributors to the development of visual-spatial skills. Some research suggests that children who experience more independence and less verbal interaction are likely to develop strong spatial skills, a result that dovetails with information obtained from anthropological studies. In a village in Kenya, children who undertook tasks that led them away from home, such as shepherding, performed better on several measures of visual-spatial ability than children remaining close to home, suggesting that children who have a wider range of environmental experiences develop richer skills.[56] Cross-cultural studies of sex-related differences in spatial functioning reveal two additional skill-learning components. Anthropologist J. W. Berry compared the abilities of Eskimos, Scots, and the Temne people of Sierra Leone, pointing out the enormous differences in visual environment they encounter. Eskimo country is open and evenly landmarked (snow covers many potential reference points), while the Temne land is covered with vegetation of various colors. The Eskimo, in order to hunt over large, relatively featureless areas, learns to be aware of minute detail. In fact, the Eskimo language is rich in words describing geometrical-spatial relationships. It is not surprising, then, that Eskimos outperform Temnes in tests of spatial ability.[57]

Child-rearing practices also differ greatly in the two cultures. Eskimos raise their children with unconditional love, only rarely resorting to physical or verbal punishment. In contrast, the Temne emphasize strict discipline, acceptance of authority, and conformity. Eskimo girls are allowed considerable autonomy, while Temne girls are raised even more strictly than the boys in this highly disciplined society. Interestingly, no sex-related differences in spatial abilities show up in the Eskimo population, although marked differences appear between Temne males and females. Berry also compared other societies, including some traditional hunting cultures, with ones undergoing Westernization.[58] In the traditional cultures there were no sex differences in spatial visualization, but differences did appear in some of the transitional ones. One hypothesis that emerges from such work is that sex-related differences in visual-spatial activities are strongest in societies in which women's social (public) roles are most limited, and that these differences tend to disappear in societies in which women have a great deal of freedom. Along these lines consider that in the United States, sex-related differences in both mathematics and spatial abilities may be changing as opportunities and roles for women change. The curricula of primary and secondary schools have become less sex-segregated with the development of equal athletic facilities and both boys and girls taking shop, typing, mechanical drawing, and home economics. As these changes continue, there is no reason to believe that sex-related differences will remain constant and every reason to assume that studies done in 1955 and in 1985 will have different outcomes.

How can we sum up some of the factors influencing the acquisition of spatial skills? Early child-parent interactions may well be involved. Plenty of studies show that parents treat boys and girls differently. Mothers are more likely to repeat or imitate vocalizations from a girl baby than from a boy baby, and they are also more likely to try to distract a male infant by dangling some object in front of him.[59] Individual personality differences also influence parent-child interactions. Preschool children have different play habits. Boys usually explore more and stay away from their parents for longer periods of time than do girls, and certainly differences in games, toys, and amount of exploration could account in part for differences in the development of spatial skills. Girls often wear physically restrictive clothing, such as frilly, starched dresses and patent leather shoes, which contributes to their more physically limited environment. As children grow older they also learn more about sex-appropriate behavior. Pressures to conform are especially strong during the teenage years, when small sex-related differences in spatial skills first consistently appear. Visual-spatial skill-dependent activities ranging from shop and mechanical drawing to mathematics and engineering are also stereotyped male strongholds, daunting to even the most talented girls. Thus the many complex components of sex-role stereotyping may be superimposed upon and may interact with earlier developmental events. In short, there is not any *one* cause of sex-related differences in visual-spatial skills. There are *many* causes. Only future research will tell which are truly significant.

The knowledge that aspects of male/female socialization very likely influence the development of male/female differences in spatial skills should not, of course, rule out the possibility that innate biological factors contribute to such differences as well. The argument I have made to this point is twofold: (1) the size of sex differences is quite small, and (2) a complex of environmental factors has *already been demonstrated* to influence the development of visual-spatial skills. Do we then even *require* the hypothesis of biologically based differences to explain our observations? I think not, although I remain open to the idea that some small fraction of an already tiny sex-related difference could result from hormonal differences between male and female.

A Plethora of Theories: Biological Storytelling

Despite the small size of sex-related differences in verbal and spatial skills, their existence has elicited numerous studies aimed at explaining them on the basis of biological differences between the sexes. Scouring the ins and outs, curves and shapes, capacities and angles of the human brain, hoping to find traits that differ in the male and female is a pastime in which scientists have engaged for more than a century. Early studies, which discovered that male brains were larger than female brains, concluded that the female's smaller size resulted in her inferior intelligence. This logic, however, ran afoul of the "elephant problem": if size were the determinant of intelligence, then elephants and whales ought to be in command. Attempts to remedy this by claiming special importance for the number obtained by dividing brain size by body weight were abandoned when it was discovered that females came out "ahead" in such measurements. The great French naturalist Georges Cuvier finally decided that intellectual ability could best be estimated by the relative proportions of the cranial to the facial bones. This idea, however, ran aground on the "bird problem," since with such a measure birds, anteaters, and bear-rats turn out to be more intelligent than humans.[60] Some brain scientists believed that the frontal lobe of the cerebrum (the part that sits in the front of the head just above the eyebrows—see figure 2.3) was an important site of perceptive powers and was less well developed in females than in males. Others argued that even individual brain cells differed in males and females, the cerebral fibers being softer, more slender, and longer in female brains.

As neuroanatomists became more and more convinced that the frontal lobe was the repository of intelligence, an increasing number of reports appeared claiming that this lobe was visibly larger and more developed in males. One report, in 1854, concluded that Woman was *Homo parietalis* (after the parietal lobe, which lies toward the back and to the side of the head—figure 2.3) and Man *Homo frontalis*. In time, however, the parietal rather than the frontal lobe gained precedence as the seat of the intellect, a change

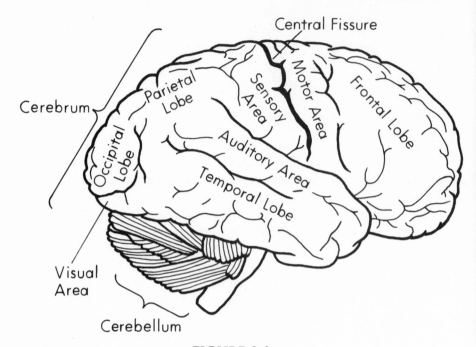

FIGURE 2.3

Cerebrum and Cerebellum

Localization of function in the human cerebral cortex. Only the major convolutions of the cortex are drawn. They are remarkably constant from individual to individual, and provide landmarks in the task of mapping the distribution of special functions in different parts of the cortex. Note especially the sensory area which lies posterior to the central fissure (or convolution), and the motor area which lies anterior to the central fissure.

accompanied by an about-face on sex differences in the brain: "The frontal region is not, as has been supposed, smaller in woman, but rather larger relatively. But the parietal lobe is somewhat smaller."[61]

Other female brain "deficiencies" found in this same period include the supposedly smaller surface area of the corpus callosum (a mass of nerve fibers that connect the left and right halves of the brain), the complexity of the convolutions of the brain, and the rate of development of the fetal cerebral cortex. These beliefs were held until 1909, when anatomist Franklin Mall used new statistical techniques developed in the budding fields of psychology and genetics to refute the existence of such differences.[62]

From the period following the end of World War I through the first half of the 1960s, psychologists and biologists developed few additional theories. A new outbreak began in the late 1960s, and since then hypotheses have come and gone rapidly. The popular press fanfares each entry with brilliant brass, bright ribbons, and lots of column space, but fails to note when each one in its turn falls into disrepute. The number and variety of theories that have come our way in the past fifteen years are truly remarkable, and an account of their advent, an analysis of their scientific basis, and a view of their demise instructive. I've listed seven of these biological hypotheses in table 2.4, along with their current status and references for studying them in more detail. The pages that follow focus attention on two of the most popular and currently active ideas— the claim that spatial ability involves a pair of X-linked genes and that male and female brains have different patterns of lateralization.

Space Genes

In 1961 Dr. R. Stafford suggested that humans carry two different X-linked genetic sites, one influencing mathematical problem-solving ability and the other affecting spatial ability.[63] Similar to Lehrke's X-linked variability hypothesis, Stafford's theory proposed that males need inherit only one X chromosome in order to excel in math or spatial tasks, while females need a math and a space gene on each X chromosome, a less frequent possibility.

If his hypothesis were true, one would expect a smaller percentage of females than of males to be good at math and spatial activities. A number of studies have tested predictions about parent-child correlations in mathematical problem solving—predictions that geneticists made from Stafford's theory. Before 1975 some small-sized studies seemed to support Stafford's contention, although the experimental results rarely obtained statistical significance (unless, in a highly unusual procedure, groups from different studies done by different research groups were pooled to increase sample size). Large studies performed since the mid-1970s have failed to find evidence to support the X-linked hypothesis. The most recent study I found concluded that "[s]ince the previous evidence from small studies cannot be replicated, it appears that the X-linkage hypothesis is no longer tenable."[64] Even more recently, Dr. Hogben Thomas, a researcher at Pennsylvania State University, pointed out that the approach used to test Stafford's hypothesis may be fundamentally flawed and that the X-linkage theory of spatial ability may simply be untestable.[65]

Furthermore, there is a very different source of data that appears to contradict the X-linked hypothesis, one recognized some years ago by two other scientists, Drs. D. R. Bock and D. Kolakowski.

TABLE 2.4
Biological Theories to Explain Sex-related Cognitive Differences

Year of Initial Publication	Name of Theorist	Basic Tenet	Current Status of Theory
1961	Stafford[g]	Spatial ability is X-linked and thus males show it more frequently than do females.	Clearly disproven,[h] although still widely quoted. Current authors still feel the necessity to argue against this genetic hypothesis.
1966	Money and Lewis[c]	High levels of prenatal androgen may increase intelligence.	Disproven by Baker and Ehrhardt in 1974.[d]
1968	Broverman et al.[a]	Males are better at "restructuring" tasks, due to lower estrogen levels, greater activity of "inhibitory" parasympathetic nervous system.	Actively critiqued in early 1970s. Not cited in current literature.[b]
1972	Buffery and Gray[f]	Female brains are more lateralized than male brains; greater lateralization interferes with spatial functions.	No evidence; not currently an important view.
1972	Levy[k]	Female brains are less lateralized than male brains, less lateralization interferes with spatial functions.	Currently in vogue; dominates the field despite a number of cogent critiques; no strong supporting evidence.
1973	Bock and Kolakowski[i]	Supplements Stafford's theory. Sex-linked spatial gene is expressed only in the presence of testosterone.	Clearly disproven,[h] although still widely quoted. Current authors still feel the necessity to argue against this genetic hypothesis.

40

TABLE 2.4 *(continued)*

1976	Hyde and Rosenberg[e]	High blood uric-acid levels increase intelligence and ambition. Males have more uric acid than females.	Not widely cited, no supporting evidence.[f]

---

[a] Donald M. Broverman, Edward L. Klaiber, Yutaka Kobayashi, and William Vogel, "Roles of Activation and Inhibition in Sex Differences in Cognitive Abilities," *Psychological Review* 75(1968):23–50.

[b] Julia A. Sherman, *Sex-Related Cognitive Differences: An Essay on Theory and Evidence* (Springfield, Ill.: Charles C Thomas, 1978); Mary Parlee, "Comments on 'Roles of Activation and Inhibition in Sex Differences in Cognitive Abilities' by Broverman et al.," *Psychological Review* 79(1972):180–84; G. Singer and R. Montgomery, "Comment on Roles of Activation and Inhibition in Sex Differences in Cognitive Abilities," *Psychological Review* 76(1969):325–27; Donald M. Broverman, Edward L. Klaiber, Yutaka Kobayashi, and William Vogel, "A reply to the 'Comment' by Singer and Montgomery on 'Roles of Activation and Inhibition in Sex Differences in Cognitive Abilities'," *Psychological Review* 76(1969):328–31.

[c] John Money and V. Lewis, "Genetics and Accelerated Growth: Adrenogenital Syndrome," *Bulletin of Johns Hopkins Hospital* 118(1966):365–73.

[d] Susan W. Baker and Anke Ehrhardt, "Prenatal Androgen, Intelligence, and Cognitive Sex Differences," in *Sex Differences in Behavior*, ed. R. C. Friedman, R. M. Richart, and R. L. Van de Wiele (New York: Wiley, 1974).

[e] J. S. Hyde and B. G. Rosenberg, *Half the Human Experience: The Psychology of Women* (Lexington, Mass.: D.C. Heath, 1976).

[f] Julia A. Sherman, *Sex-Related Cognitive Differences: An Essay on Theory and Evidence* (Springfield, Ill.: Charles C Thomas, 1978).

[g] R. E. Stafford, "Sex Differences in Spatial Visualization as Evidence of Sex-Linked Inheritance," *Perceptual and Motor Skills* 13(1961):428.

[h] Robin P. Corley, J. C. DeFries, A. R. Kuse, and Steven G. Vandenberg, "Familial Resemblance for the Identical Blocks Test of Spatial Ability: No Evidence of X Linkage," *Behavior Genetics* 10(1980):211–15.

[i] D. R. Bock and D. Kolakowski, "Further Evidence of Sex-Linked Major-Gene Influence on Human Spatial Visualizing Ability," *American Journal of Human Genetics* 25(1973):1–14.

[j] A. W. H. Buffery and J. Gray, "Sex Differences in the Development of Spacial and Linguistic Skills," in *Gender Differences: Their Ontogeny and Significance*, ed. C. Ounsted and D. C. Taylor (London: Chirhill Livingston, 1972).

[k] Jerre Levy, "Lateral Specialization of the Human Brain: Behavioral Manifestation and Possible Evolutionary Basis," in *The Biology of Behavior*, ed. J. A. Kiger Corvalis (Eugene: University of Oregon Press, 1972).

Rather than discard Stafford's hypothesis, however, they modified it, turning counter-evidence into support.[66] On occasion, individuals are born with no Y chromosome. Doctors call them XOs. Since they are born with female genitalia, XO individuals are usually raised as girls, and in many respects are quite normal, although they can sometimes be recognized by their short height, webbed neck, and failure to develop fully at puberty. XO individuals, said to have Turner's Syndrome (named after the physician who first described it), have spatial abilities well below the normal range, a fact that contradicts Stafford's hypothesis. If the X-linked hypothesis were correct, Turner's Syndrome patients would not differ from XY males, expressing their spatial ability more frequently than XX females, because their single X chromosome is not "covered" by a second X. In order to get around this uncomfortable fact, Bock and

TABLE 2.5

Verbal and Performance IQ's of Individuals with Sex Chromosome and/or Hormone Abnormalities

Number of Individuals Tested	Sex of Rearing	Sex Chromo- some; Constitution	Adult Hormone Levels	Average Verbal IQ Scores	Average Perfor- mance IQ Scores
45	F	XO	low estrogen low androgen	106	86
15	F	XY	intermediate estrogen, androgen- insensitivity	112	102
3	M	XY	intermediate estrogen, androgen- insensitivity	117	119
23	M	XXY	intermediate estrogen, intermediate androgens	105	88
12	M	XXY	intermediate estrogen, intermediate androgens	66	76
20	M	XYY	unknown	79	88

NOTE: Julia Sherman, *Sex-Related Cognitive Differences: An Essay on Theory and Evidence* (Springfield, Ill.: Charles C Thomas, 1978), 84. Courtesy of Charles C Thomas, Publisher.

Kolakowski proposed that the space gene is not only X-linked but is also sex-limited, depending for its expression on high androgen levels which circulate throughout the body in higher concentrations in men than in women. (A familiar example of a sex-limited gene is baldness, expressed only in men because it depends for its expression on higher androgen levels than are present in most females.)

The sex-limited hypothesis represents a clever stab at saving the game, but it too runs counter to the data. Psychologist Julia Sherman has offered the most succinct demolition of the theory, and table 2.5 represents some of her work.[67] Turner's Syndrome patients have lower than normal estrogen (a hormone found in higher concentrations in females) and androgen levels. Bock and

Kolakowski argue that the gene coding for spatial ability requires a certain cellular concentration of androgen in order to function. In XO individuals, they suggest, too little androgen is present, and thus Turner's Syndrome girls have poor spatial abilities. To shore up their position, they cite another study of individuals with androgen insensitivity syndrome (AIS)—people who possess both X and Y chromosomes but who are unable to respond to androgens. AIS patients are often born with femalelike genitalia. Fifteen such persons, all raised as females, were tested and obtained an average Verbal IQ of 112 and Performance IQ of 102.* Although both scores fit in the normal range, Bock and Kolakowski inferred from this test that inability to respond to androgen lowered spatial IQ. But who can say whether the Verbal IQ might not have been abnormally high rather than the spatial IQ being unusually low? Furthermore, Bock and Kolakowski ignore additional data from the same study. Three AIS patients reared as *males* scored well above the normal range on both verbal and spatial IQ tests. If androgen really improves the expression of spatial genes, how is it that three androgen-insensitive individuals performed above average on a spatial test?[68]

Chromosomal abnormalities affect mental functioning. All people born with either one too many or one too few chromosomes show some degree of mental impairment. The information in table 2.5 makes this clear. Only AIS patients, who have a normal chromosome complement, score consistently in the normal range on both Verbal and Performance IQ. The data in table 2.5 thus suggest that good performance correlates with normal chromosome complements, *not*—as Bock and Kolakowski suggest—with hormone levels. By any scientifically acceptable standards, this attempt to save the X-linked space gene theory fails.

As a study in the sociology of science, however, the Stafford hypothesis remains interesting. From the point of view of a geneticist, the idea that two specific genes govern a complex, continuously varying trait is dubious to begin with. As we have just seen, the available data is either categorically inappropriate or lends no support to the idea. Yet since its initial publication in 1961, the X-linkage hypothesis has shown considerable tenacity, appearing as fact in some textbooks and showing up in highly political articles as part of larger arguments about the genetic incapacity of females for certain sorts of work.[69] The real fact is that many people, both

---

* Performance IQ (see p. 32) is used by some scientists as a measure of spatial ability, although it is not a test designed for this use.

cathy®   by Cathy Guisewite

SEE WHAT YOU CAN DO WITH THE FALLON PROJECT, CATHY.

RUSTLE RUSTLE STAPLE CLIP RUSTLE

CATHY

MY RIGHT BRAIN TOOK OVER.

Copyright © 1984, Universal Press Syndicate. Reprinted with permission. All rights reserved.

scientists and nonscientists, just plain *like* the idea and go to considerable lengths to salvage it because it fits so neatly into the entrenched stereotype of feminine inferiority. It constitutes a not uncommon example of how social views influence the progress of science.

Left versus Right: The Psychologists' Sleight of Hand

Functionally, humans have two brains. The idea has become sufficiently commonplace to appear even in the daily newspaper cartoons (see cartoon, above). While the left hemisphere of the brain appears specialized to carry out analysis, computation, and sequential tasks, in the right half resides artistic abilities and an emotional, nonanalytic approach to the world. As originally developed, the idea of brain hemisphere differentiation said nothing about sex differences. But it didn't take long for some scientists to suggest that left–right brain hemisphere specialization could "explain" supposed male/female differences in verbal, spatial, and mathematical ability. The development, dissemination, and widespread acceptance of such ideas provides a second and still very active example of science as social policy.

Humans, like all vertebrates, are bilaterally symmetrical. Although our left and right sides represent approximate anatomical mirror images of one another, they are not equally competent at the many daily activities in which we engage. Each of us has a particular hand and foot preference, using one side of the body more skillfully than the other to, among other things, kick a football, throw a baseball, write, or eat. Such functional asymmetry provides one tangible measure of a complex and poorly understood division of labor between the two sides of the brain. Looking down on the brain from above, one sees the convoluted folds of the right and left halves of the cerebral cortex connected by an enormous mass of nerve fibers, the corpus callosum (see figure 2.4). Each brain hemisphere controls movements executed by the opposite side of the body. Most people are right-sided, that is, they perform most major activities with the right side of the body, and can thus also be thought of as left-brained. The common scientific belief is that the left hemisphere controls the right side of the body's activities. The converse is probably true for many but not all left-siders.

Our understanding of how the brain mediates our behavior remains superficial, yet a few general observations are possible. For starters, we know that different portions of the cerebral cortex have primary responsibility for particular functions. For example, a region in the posterior part of the cortex (the part located at the back of the head, just above where the skull and neck hook together) enables us to see (and is thus referred to as the *visual cortex*). The region of the cortex responsible for hearing is located further forward along the left side of the head, and numerous other functions take up primary residence in other regions of the brain, as figure 2.4 illustrates.

A second aspect of brain function involves the notion of cerebral dominance. For many years scientists thought of the hemisphere controlling our preferred side as the major hemisphere, and the other as a minor, less competent half. During the 1950s a change in that viewpoint evolved, because discoveries made it clear that the halves of the brain were not so much dominant and dominated

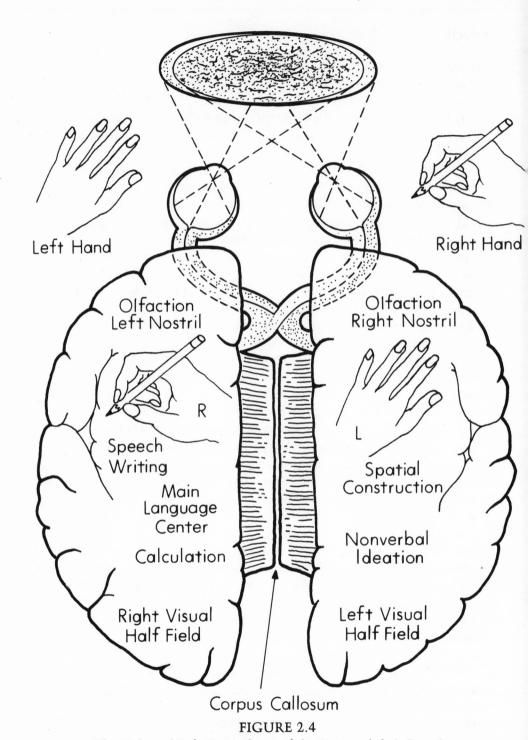

FIGURE 2.4

The Right and Left Hemispheres of the Brain and their Functions

NOTE: Richard Restak, *The Brain, The Last Frontier* (New York: Warner, 1979), 190. Copyright © 1979 by Richard M. Restak. Reproduced by permission of Doubleday & Company, Inc.

as they were different. Some physicians tried, with a modicum of success, to control severe cases of epilepsy by cutting the fibers of the corpus callosum, thus separating the connections between the two halves of the brain. Because patients receiving such operations appeared under most circumstances to function normally, some time passed before Dr. Roger Sperry, a well-known neurobiologist, and a number of his students designed special tests which revealed that in such "split-brain" individuals the different halves of a supposedly symmetrical brain had very different capabilities.[70] They found, for example, that if a blindfolded split-brain patient picks up a pencil in the left hand (controlled by the right brain) he or she can neither name nor describe it, although such a patient can easily do so while holding the pencil in the right hand. Such an observation suggests that language—the ability to read and speak—is localized primarily on the left side of the brain. The fact that the same patient (still blindfolded) can use his or her sense of touch to select a pencil from among a number of other objects shows that the ability to recognize pencils remains. But when the fibers connecting the two halves of the brain are severed, there is no transfer of this recognition from the right side of the brain to the left, where the ability lies to name what the patient touches. The same is true of visual perception. With the right eye covered, these patients cannot read or copy words flashed in front of the left eye (connects to the right hemisphere), although they recognize the content. One of Sperry's patients, for example, gave an embarrassed giggle when a nude figure flashed before her left eye, but she could not explain why she had laughed.

As a result of many studies on both split-brain patients and people in whom one brain half has been damaged by a stroke, cancer, or accident, scientists now make the following generalizations about normal, *right-handed* people. (Little is yet understood about left-sidedness.) The left side of the brain has the capability for all verbal activities, and for analytical, mathematical, and sequential information processing. It is sometimes called the analytical brain. The right side specializes in spatial skills and holistic, nonverbal, Gestalt processing, including musical ability. A concert conductor who, due to a stroke, was almost completely unable to speak but could nevertheless continue to conduct his own orchestra, provides a dramatic illustration of this hemispheric specialization.

The fact that the two halves of the brain specialize for different intellectual activities is of both theoretical and practical importance.

47

Humans are the only primates exhibiting handedness or hemispheric specialization,[71] and some speculate that the phenomenon may have evolved as part of the evolution of speech and tool making. But consideration of the discoveries that have located certain functions in particular regions of the brain requires some caution. We have yet to understand how the brain thinks, and we know nothing about how, or even whether, the brains of two individuals—one skilled in mathematics and the other a talented fiction writer, for example—differ. To illustrate, fantasize for a moment about the return to earth of Benjamin Franklin, the inventor, scientist, and patriot. The year is 1985. On his return Franklin observes immediately that our roads crawl with mechanical horses (cars). Curious, he experiments to discover how they work. First he removes different parts of the car, observing that the removal of a wheel makes the ride bumpy, draining of the brake fluid makes it difficult to stop, and removal of the battery or engine prevents forward motion altogether. Although he thus localizes some functions, he uncovers little information about their mechanisms or about the natures of either batteries or internal combustion engines. Only much more painstaking analyses could reveal that. Ben Franklin reincarnate might quickly identify the seat of motive power, but he could not as easily uncover its mechanism of function.

In finding that different halves of the cerebrum specialize for different functions, we have identified a major seat of power but have learned little about how it operates. A 1980 article in *Science* magazine further illustrates how little we know. It contains a report on a British university student with an IQ of 126 who has first-class honors in math, is socially normal, yet has hardly any nerve cells in his cerebral cortex.[72] This and similar medical reports suggest that the task of understanding the mechanisms by which the brain performs intellectual functions still lies far beyond our reach.

The excitement elicited by Sperry's discoveries has led to a somewhat unbalanced view of how the brain works. As Sperry himself comments in his Nobel Laureate address: "The left-right dichotomy . . . is an idea with which it is very easy to run wild."[73] He cautions that other divisions in the brain (such as front/back, up/down) may also have unrecognized importance but, most important, he stresses that the brain operates as a coherent whole, a closely integrated unit. Overemphasis on the separate abilities of

48

particular brain regions easily leads us to neglect to inquire into the function of the integrated whole.

Sperry also suggests that each human brain may be different enough to defy generalization: "The more we learn, the more we recognize the unique complexity of any one individual intellect and the stronger the conclusion becomes that the individuality inherent in our brain networks makes that of fingerprints or facial features gross and simple by comparison."[74] Since scientists work on the assumption that they can generalize and predict, Sperry's suggestion is quite unsettling. If he is right, then entire subfields of psychology and neurobiology may have to change their approach and their focus.

One last point about hemispheric specialization: cerebral lateralization is not immutable during childhood. In children who incur brain damage on only one side, the undamaged hemisphere can carry out all of the functions of an uninjured brain, although in adults this is not the case. It seems at least plausible, then, that the developmental environment of childhood plays an important role in the attainment of adult hemispheric capacities.[75]

Not long after the discovery of hemispheric specialization, some scientists began using it to explain both the supposed female excellence in verbal tasks and the male skill in spatial visualization. In the past eight years at least four different theories on these skills have appeared, the two discussed here having received the most attention although, interestingly enough, they are mutually incompatible. The first, put forth in 1972 by two psychologists, Drs. Anthony Buffery and Jeffrey Gray,[76] now suffers disfavor. The other, elaborated by Dr. Jerre Levy[77]—who during and after her time as Sperry's student, played an important role in defining the modern concept of hemispheric specialization—is still in fashion. The pages of *Psychology Today*, *Quest*, and even *Mainliner* magazine (the United Airlines monthly) have all enthusiastically described her theory. Speculation also abounds that sex differences in hemispheric specialization result from different prenatal and pubertal hormonal environments.[78] Since a number of psychologists have pointed to a substantial body of experimental evidence that renders Buffery and Gray's hypothesis untenable,[79] we will consider only Levy's views.

Levy hypothesizes that the most efficiently functioning brains have the most complete hemispheric division of labor.[80] Women, she suggests, retain a capacity for verbal tasks in both hemispheres.

In other words, they are less lateralized for speech than are men. When verbal tasks in women "spill over" to the right side of the brain, they interfere with the right hemisphere's ability to perform spatial tasks.* Men, in contrast, have highly specialized brain halves—the left side confining its activities solely to verbal problems, the right side solely to spatial ones.

Let's suppose for a moment that male and female brains do lateralize differently and ask what evidence exists to suggest that such differences might lead to variations in performance of spatial and verbal tasks. The answer is, quite simply, none whatsoever. Levy derives the idea not from any experimental data but from a logical supposition. In her later work[81] she takes that supposition and "reasons" that "a bilaterally symmetric brain would be limited to verbal or spatial processing. . . ." Recently psychologist Meredith Kimball reviewed the small number of studies that might act as tests of Levy's logical supposition and came up empty-handed, concluding that there is no evidence to support the key assumption on which Levy builds her hypothesis.[82]

Nevertheless, the proposal that men and women have different patterns of brain lateralization has provoked enormous interest. Scientists have published hundreds of studies, some done on normal subjects and others derived from subjects with brain damage due to stroke, surgery, or accident. The idea that verbal function might operate differently in male and female brains came in part from a long-standing observation: among stroke victims there appear to be more men than women with speech defects serious enough to warrant therapy. There may be a number of explanations for why men seek speech therapy more frequently than do women. To begin with, more males *have* strokes.[83] Also, it is possible that males seek remedial therapy after a stroke more frequently than do females. And strokes may affect speech less severely in females because females have better verbal abilities *before* the illness.[84]

Some researchers have attempted to sort out these possibilities, but a controlled study of stroke victims is extremely difficult. One reason is that there is no way of knowing for sure whether male and female victims under comparison experienced exactly the same type of brain damage. Even comparisons of individuals who had surgery performed on similar parts of their brains are probably

---

* The theory actually holds that left-handed men resemble women in this regard. The experimental support for her conclusions about left-handed men has been roundly criticized, especially by J. Marshall (see note 79).

quite misleading because of variation in brain morphology from individual to individual. It would be possible to ascertain the exact regions of the brain affected only by looking at microscopic sections of it, a practice that is routine in animal experiments but would of course be impossible with live human beings. Extensive reviews of clinical studies reveal a great deal of controversy about their meaning, but little in the way of strong evidence to support the idea that women have bilateralized verbal functions.[85] Consider the statement of a scientist who believes her work to *support* the differential lateralization hypothesis:

> Neither do the data overwhelmingly confirm that male brains show greater functional asymmetry than female brains. . . . One must not overlook perhaps the most obvious conclusion, which is that basic patterns of male and female brain asymmetry seem to be more similar than they are different.[86]

If this is the kind of support the proponents of sex differences in laterality put forward, then it is amazing indeed that the search for sex-related differences in brain lateralization remains such a central focus of current research in sex-related cognitive differences.

In addition to looking at patients with brain damage, researchers have tested Levy's hypothesis using normal individuals. The most common way of measuring hemispheric specialization in healthy people is by the dichotic listening test. To look for language dominance, experimenters ask the subject to don a set of headphones. In one ear the subject hears a list of numbers, while in the other he or she simultaneously hears a second, different list. After hearing the two lists, the subject (if not driven nuts) must remember as many of the numbers as possible. Usually subjects can recall the numbers heard on one side better than those heard on the other. Some experimenters believe that right-ear excellence suggests left-hemisphere dominance for verbal abilities and vice versa, but this conclusion ignores other possibilities. Individuals who take the tests may develop different strategies, for instance, deciding to try to listen to both sets of numbers or to ignore one side in order to listen more closely to the other.

Some scientists have reported sex differences in performance on dichotic listening tests, but three recent reviews of the research literature indicate a lack of solid information.[87] Many studies show no sex differences and, in order to show any differences at all, large samples must be used, all of which suggests that same-sex disparities may be larger than those between the sexes. One reviewer ends her article with the following comments:

> Any conclusions rest on one's choice of which studies to emphasize and which to ignore. It is very tempting to . . . argue that there are no convincing data for sex-related differences in cognition or cerebral lateralization. . . . In fact, what is required is better research.[88]

Analogous methods exist for studying visual lateralization. Tests utilize a gadget called a tachistoscope, through which a subject looks into a machine with an illuminated field. The machine flashes different items in front of either the right or the left eye, and the subject tries to identify as many as possible. Nonverbal images such as dots (as opposed to words or letters) suggest some left-field (right-hemisphere) advantages for men, but here too the data vary a great deal. For example, many (but not all) studies show male left-eye advantages for perception of photographed faces, scattered dots, and line orientations, but no sex differences for the perception of schematic faces, depth, or color.[89] In addition, the fundamental question of whether such tests have anything at all to do with brain lateralization continues to cloud the picture.

Although there is no solid evidence for the idea that females are more bilateral than males in verbal functioning, there *does* seem to be evidence that females use their left hemisphere (their verbal hemisphere) more frequently to solve visual-spatial problems. As Sherman points out, this does not necessarily imply a sex-related difference in brain organization, but could instead reflect different problem-solving strategies. For whatever reasons, females may prefer to use verbal approaches to the solution of spatial problems. In fact, several studies show different *approaches* to nonverbal problem solving, but find no overall sex differences in performance. In other words, males did not have a better final outcome; they just reached the same end using a different means than did the females.[90]

As of this writing, a number of hypotheses to explain such strategic differences are actively competing with one another. Sherman, for example, suggests what she calls the "bent twig" hypothesis,[91] proposing that girls develop their language ability a bit earlier than do boys, thereby initiating a chain of reactions that give females progressively greater language skills. Because girls talk sooner, parents may talk more with and further develop their daughters' language skills. And because of their facility, little girls may choose verbal mediation over so-called "Gestalt" processes for the solution of visual-spatial problems. Sherman entertains the possibility that early verbal development in young girls is a true biological sex difference, but it is also possible that if girls learn to talk earlier than do boys it is because adults in their environment interact with them more often and intensively with the spoken word.

If there remains uncertainty about different maturation rates in early language acquisition, there is some clarity in the fact that on the average, girls reach puberty and adult size two to three years before boys. This developmental rate difference forms the basis of another hypothesis, put forth by psychologist Dr. Deborah Waber.[92] She provides evidence that late maturers, *male or female,* have more highly lateralized brains. She thus accounts for any small male/female differences in lateralization by the fact that males, on the average, grow more slowly. Not all studies, however, agree either about the correlation of maturation rate and spatial abilities or about the interpretation of any such correlation, and Waber's suggestion remains under active investigation.[93]

Finally, there is a series of suggestions, not yet in the form of full-dress hypotheses, about the ways in which physical activity might affect the development of visual-spatial skills. What kinds of cognitive capacities develop from active play—tree climbing, running, throwing, batting, and catching a ball? Virtually no information exists on this issue. Yet if scientists are truly interested in how cognitive abilities develop (in a little boy or a little girl), these questions surely require investigation.

Biological Calculus: Do the Sexes Differ in Mathematical Ability?

A few years ago, a friend phoned me for some advice. His ten-year-old daughter was upset because she had just heard on the radio about the hot new discovery that boys are genetically better at math than are girls. Girls, she had heard, would be less frustrated if they recognized their limits and stopped their fruitless struggle to exceed them.

"Daddy," she had said, "I always wanted to be a math professor like you. Does this mean I can't?"

My friend wanted to know if I had read the article. "Is it true? What can I tell my daughter?"

Just two days before, I had seen the same report in the *New York Times*.[94] One day before, the mail carrier had dropped through my mail slot the issue of *Science* magazine containing the short research article by Drs. Camilla Benbow and Julian Stanley, which I had seen summarized in the *Times*.[95] Within the week, radio advertisers hawked the latest issues of *Time* and *Newsweek*, telling me even as I sleepily brushed my teeth in the morning to buy the magazines because they contained new evidence about "male math genes."[96] And so it went. The *Time* article even had an illustration in case we couldn't get the written message. The cartoon portrayed a girl and a boy standing in front of a blackboard, with a proud, smug-looking adult—presumably a teacher—looking on. The girl frowns in puzzlement as she looks directly out at the reader. On the blackboard in front of her stands the multiplication problem 8×7, which she is clearly unable to solve. The boy looks with a toothy smile toward the adult, who gazes back at him. The cause for the satisfaction? The correct answer to the multiplication problem $7,683 \times 632$. Interpreting the image does not require a degree in art history, and the aftershocks from the *Science* article and subsequent press coverage still rumble beneath our feet.

Clearly, math and sex is a hot topic. The question is, what's all the fuss about? The issue is part of both a national and individual crisis. The 1983 report of the National Science Board's Commission on Precollege Education in Mathematics, Science and Technology puts it this way: "Our children could be stragglers in a world of technology. We must not let this happen; America must not become an industrial dinosaur. We must not provide our children a 1960s education for a twenty-first-century world."[97]

Researcher Lucy Sells labels high school mathematics achievement a "critical filter" which limits the choices of study available to women and minorities who enter college,[98] and others have pointed out how mathematics requirements restrict entry to the high-paid field of engineering.[99] A recent counseling session I had with a female student interested in biology vividly illustrates the point. The student wanted advice about choosing her major. She had taken a variety of nonscience courses during her first couple of years in college and was certain that she did not want to become a high-powered professional, preferring instead a goal of working one-on-one with people in city neighborhoods. "And," she said, "I really like biology." Fresh from reading the National Science Board's report, I cheerfully suggested that she might translate her enthusiasm for this aspect of science into becoming a science and math teacher at the primary or secondary level. It was then that her face fell. "I haven't had math since my sophomore year in high school," she confessed, "and that seems so long ago, I can't even remember whether it was a course in algebra or geometry."

That left us stuck. I urged her to take the preintroductory level course offered by our math department (pejoratively called "math for poets" by some), with the hope that she could build up enough background to enter the introductory sequence. But I realized that by the time she had done that she would be ready to graduate. The choice she had made to stop studying math in high school now, five years later, had come back to haunt her. The damage was not irremediable but it would take time, and I would be surprised if she ended up deciding to teach math and science, badly needed though she may be.

There are very few hard facts about women and mathematics, but one thing we know for sure is that girls take fewer mathematics courses in high school than do boys.[100] Do they drop out because they are less able to achieve mathematically, or are other factors involved, such as stereotypes about math being a male field, discouragement from parents and teachers, and social pressures? Most educational researchers agree that in an unselected population, boys and girls are equally good at math until the seventh grade. Beyond that, however, the debate becomes mired in confusion. Many studies done before 1974 that claimed to find significant sex-related differences in math achievement failed to use well-matched populations. Instead the boys averaged a larger number of courses taken, and the studies really compared girls who had taken only one or two math courses with boys who had taken three or four. Later work that attempted to control for both the number of math courses and the number of related courses in areas such as mechanical drawing and drafting sometimes found only small sex-related differences favoring boys.[101] The results of studies done on large unselected samples give inconsistent results.[102]

In some of the studies for which sex-related differences in math have been found even among students with the same number of formal math courses, the role of social factors in accounting for such differences has also been measured. Differences in spatial visualization seem to play only a minor role, but other factors, such as the perceived importance of mathematics for future studies (girls less often thought it important), the perception of math as a male field (more male engineers and math professors), active discouragement of girls by teachers and parents (girls more often received negative feedback than boys), all taken together went a long way toward accounting for the small, occasional differences found in mathematics achievement between boys and girls who have taken the same number of math courses.

This being the case, how is it that the Benbow-Stanley report in *Science* provoked such widespread publicity? The authors make use of data obtained from ongoing studies of mathematically precocious youth at Johns Hopkins University. Since 1972 these researchers have run talent searches to find seventh and eighth graders who are unusually good at math, generally students found to perform in the upper 3 to 5 percent of their classes. Of the talented youth identified, 43 percent were girls. As part of their study, Benbow and Stanley gave these seventh and eighth graders the College Board Scholastic Aptitude Test in mathematics. Since such tests are designed for juniors and seniors in high school who are older and who have had more advanced math classes, Benbow and Stanley argued that the exams functioned as achievement tests in mathematical reasoning when given to much younger students. In each of the six years of the study, the results showed that on average girls obtained scores that were between 7 and 15 percent lower than the average score for boys. Furthermore, anywhere from three to ten times more boys than girls score in very high ranges, although there were usually some very high scoring girls. Such are the results; the fight, of course, is about their meaning. Benbow and Stanley point out that in the seventh and eighth grade, most children take the same math courses, so the differences cannot be due to *formal* course taking:

> It, therefore, cannot be argued that these boys received substantially more formal practice in mathematics and therefore scored better. Instead, it is more likely that mathematical reasoning ability influences differential course-taking in mathematics.

After further considering their data, they end their article with the following.

> We favor the hypothesis that sex differences in achievement in and attitude toward mathematics result from superior male mathematical ability. . . . This male superiority is probably an expression of both endogenous and exogenous variables . . . the hypothesis of differential course-taking was not supported. It also seems likely that putting one's faith in boy-versus-girl socialization processes as the only permissible explanation of the sex difference in mathematics is premature.[103]

In a subsequent paper, also published in *Science*, the same researchers added more subjects to their data base, a fact which they believe further substantiates their initial conclusion.[104]

Attacks on Benbow and Stanley's conclusions have been of three types: questions about the validity of the aptitude test, questions about the limitations of looking only at formal mathematical experience, and questions about whether boys and girls receive the same training even within the same math course. The debate over achievement versus aptitude tests is complex and, I feel, a little beside the point.[105] Whether one calls it aptitude, achievement, or reasoning ability, the fact remains that more very bright boys performed exceptionally on Benbow and Stanley's tests than did very bright girls. Benbow and Stanley claim they are doing nothing more than telling it like it is: "It is not the method of science . . . to ignore published facts or provide a forum for subjective judgments and anecdotal evidence."[106] Since they seem willing to take on the Galilean stance of the honest but persecuted scientist in this debate, it is surprising that they fail to cite a number of studies—including both some of their own and ones by other workers in the Johns Hopkins project—that would at the very least dilute the emphasis they place on the idea of "superior male mathematical ability."

For example, they specifically attack psychologists Elizabeth Fennema and Julia Sherman's conclusion about differential course taking; yet those very same researchers in the very same articles write at length about a variety of other factors (parental attitudes, teachers' attitudes, informal mathematics experience), *all* of which contribute significantly to mathematical achievement. Another group of researchers observed thirty-three second-grade teachers as they taught reading and arithmetic in the classroom, finding that teachers spent more classroom time teaching reading to individual girls and less teaching them math. The boys received less direct instruction in reading and more in math.[107] In other words, boys and girls learning together in the same classroom did not receive the same instruction. Benbow and Stanley dismiss this study as irrelevant, saying only that it seemed inapplicable to studies of highly talented children.

Furthermore, in their most recent work they cite some of their own data[108] to support their skepticism about the inability of socialization to account for the reported achievement test differences in talented junior high school students. Surprisingly, though, they dismiss as irrelevant other information they have collected showing that (1) in high school the girls in their study get slightly better grades in math courses than do the boys,* and (2) when asked about intended college majors: "The percentage of males reporting that they intended to major in the mathematical sciences was 15%, while for females this was 17%."[110]

Dr. Helen Astin looked at the backgrounds of children in the early years of the Johns Hopkins Study and found that "parents of boys admit that they encouraged the boys more by giving them science kits, telescopes, microscopes, or other science-related gifts."[111] In addition, the parents of mathematically precocious boys had higher educational hopes for them than did the girls' parents. Finally, Lynn Fox and Sanford Cohn, also researchers in the Johns Hopkins Study, find evidence from other aspects of the project to "support the social explanation of sex differences at higher levels of ability and achievement."[112] The evidence from these talented youngsters that boys and girls even at the seventh-grade level have had different experiences with regard to math training and have developed very different hopes for their futures makes implausible Benbow and Stanley's underlying assumption that only socialization events at puberty influence the development of mathematical skill and achievement.

In considering the debate over math ability the question foremost in my mind is, Why the rush to judgment? Benbow and Stanley, although they hedge their bets, clearly assert that boys have greater math ability than do girls. While choosing to simplify the discussion and ignore much of the literature on sex differences in both teaching and informal learning, as well as findings on career goals and hopes, they invite us all to "face facts" and accept the biological nature of sex differences in math ability if that is, indeed, what objective science proves.

* When criticized for failing to discuss the discrepancy between achievement in the classroom and scores on SAT tests, Benbow and Stanley dismissed their own finding by attributing it "to the better conduct of girls in school."[109]

We *do* know a great deal about mathematics learning and especially about why girls drop out of math classes.[113] If math and science were required for four years of high school, if girls were actively counseled to consider careers in science and warned about the ways in which dropping out of math limits their future choices, if teachers were made aware of the different ways they treat boys and girls in the classroom, and if there were many more women teaching math and science to our youngsters, the "problem" of women in math would lessen dramatically, and in all likelihood would disappear. To argue for "endogenous" differences, as Benbow and Stanley have, is to argue willy-nilly, not to bother with all of the "exogenous" changes in educational method and quality that we could make with some degree of success right now. At best their call to consider "natural" causes for the sex difference in math is premature. If, once we have reformed our informal and formal systems of mathematics education and career counseling, there remain significant sex differences in mathematics ability, *then* might be the time to wonder about innate sex differences. In the meantime there are a lot of very specific changes to be made in how we educate our young people. What, then, are we waiting for?

A Question of Genius: Some Conclusions

Are men really smarter than women? The straightforward answer would have to be no. Early in this century, scientists argued that there might be more male than female geniuses because male intelligence varied to a greater extent than did female intelligence. This "fact" provided proof positive of the overall superiority of the male mind. Hypotheses in defense of this position still pop up from time to time. They consist of old ideas in modern dress and are unacceptable to most mainstream psychologists. In apparent contrast Maccoby and Jacklin believe that males and females are equally intelligent while entertaining the possibility that the two sexes have somewhat different cognitive skills; they suggest a biological origin for such differences. Although the possibility is admissible, I have tried to show both that any such differences are very small and that there is no basis for assuming a priori that these small variations have innate biological origins.

This chapter bears witness to the extensive yet futile attempts to derive biological explanations for alleged sex differences in cognition. Although these efforts all have a certain social wrong-headedness to them, they do not stand or fall on their political implications. Rather, such biological explanations fail because they base themselves on an inaccurate understanding of biology's role in human development. Sperry suggests this when he writes that each person's brain may have more physical individuality than do the person's fingerprints. His statement is radical because it implies that attempts to lump people together according to broad categories such as sex or race are doomed to failure. They both oversimplify biological development and downplay the interactions between an organism and its environment. As a result of doing the research for this book, I arrived at the same conclusion. My feelings come from having thought carefully about the present state of our knowledge about the genetics of behavior, the embryological development of the sexes, and the ways in which hormones act as physiological controllers and evocators in males and females. By coming to understand these aspects of human development, we can see more clearly why simple, unidirectional models of biological control of human behavior misconstrue the facts of biology. We will explore this phenomenon in the next chapter.

3

OF GENES AND GENDER

Recent research has established beyond a doubt that males and females are born with a different set of "instructions" built into their genetic code. Science is thus confirming what poets and parents have long taken for granted.
—TIM HACKLER
Mainliner, 1980

The relation between gene, environment and organism is not one-to-one but many-to-many. Given the genes and the environment one cannot predict the organism. Given the organism, we cannot infer its genotype or the environment in which it developed.
—RICHARD LEWONTIN
Human Diversity, 1982

LET'S FACE IT: the belief that genes dictate our behavior has enormous appeal. For starters, our daily observations seem to confirm the notion. The other day, for example, I came upon a photograph of myself, aged nine, and realized with astonishment how much I looked like my nine-year-old niece. Not only do we regularly observe physical similarities among family members, we also note behavioral ones. If genes account for the fact that both my mother and I have curly hair, then might they not also explain why my father, brother, and I all gesture energetically when we speak? Family resemblances, be they genetic or learned, offer deep psychological comfort. My amazed recognition of myself physically reincarnated in the person of my niece matched my pleasure at sensing immortality. My body will some day pass from this earth, but part of me will remain, passing itself on, perhaps, to generations yet unborn. And should my children turn out to be brilliant or successful, so much the better. I can claim half the credit, since half their genes came from me.

But genes are double-edged. One invokes them to take praise for the good, but blames them to get emotional respite from the bad. (My weight problem? It's not *my* fault. I inherited a craving for sweets; besides, the women in my family tend to be overweight.) We can even blame our genes for social ills. Anthropologist Lionel Tiger, for instance, writes that failure to accept women's genetic limitations leads to economic discrimination. Implying that political efforts to obtain male/female equality seem likely to fail, he suggests instead that social scientists should reconsider their nongenetic view of the origins of widespread social patterns but "particularly the effect of genes on the situation of women."[1]

The popular press and scientists alike have apparently fallen in love with the gene. Maya Pines, for example, wrote in the *New York Times* about the rapidly expanding field of behavior genetics. Not only have scientists in this area begun to talk about things such as the genetic basis of shyness, she writes, in addition they "are establishing ties between heredity and stuttering, dyslexia and alcoholism."[2] At the same time, reports continue to appear, even in very recent scientific publications, linking genes to criminal behavior[3] and, as always, to IQ.[4]

Moving from physical similarities among family members to genetic bases for individual behavior such as shyness, alcoholism, or criminality and then beyond, to consider genetic causes of complex social structures, we jump over enormous intellectual chasms. In the following pages we will look before we leap at what lies below, examining the questions: what are genes and how do they work; how do genes interact with environment to affect behavior; and how do genes participate in the genesis of biological differences between males and females? In considering these topics we must keep in mind the seductive nature of genetic explanations, allowing ourselves to sustain some intellectual discomfort in search of solid and accurate explanation. The strong—albeit erroneous—belief that genetic differences are unchangeable make this neither an easy nor an idle task. More often than not the phrase *genetic basis* encodes the meaning "change is impossible."[5]

What Are Genes and How Do They Work?

In the beginning scientists had no physical account of the gene. Gregor Mendel, for instance, abstracted a set of rules to describe the inheritance of traits in adult pea plants, such as red or white flowers or smooth or wrinkled peas. His rules, which grew out of observations of the final stage of plant development (rather than, for example, from looking at the pea plant embryo), depended upon the frequency of appearance of a particular trait, or *phenotype*, as geneticists label characteristics such as flower color, fruit texture, plant height, and so on, which make up the appearance, behavior, and growth characteristics of the plant. Mendel described something in peas that we all are familiar with in humans. A brown-haired set of parents may have only brown-haired children, but those children will sometimes produce blond offspring. In Mendelian terms we describe this observation by speaking of factors, which the early Mendelians called genes, which lead to the production of different hair colors. These factors exist in alternate states (for example, red rather than white flowers, smooth versus wrinkled fruit, or blond versus brown hair) called *alleles*, which influence the development of a particular phenotype.

Mendel abstracted from his results the rule that some alleles dominate others. A *dominant* allele for brown hair (designated *B*), for example, when paired with a *recessive* allele for blond (designated *b*) results in the appearance of a brown-haired *Bb* child. The pairing of two recessive alleles would result in a blond (*bb*) child. Since two different gene combinations, *BB* and *Bb*, can produce a brunette, geneticists say that two different *genotypes* can produce the same *phenotype*. *BB* and *bb* individuals are called *homozygous* and *Bb* individuals are *heterozygous*. A shorthand conceptualization of these possibilities as applied to round and wrinkled peas appears in figure 3.1.

The distance traveled from Mendel's abstract rules of inheritance to our present-day conceptualization of genetic information is great. In the first quarter of this century many geneticists pictured genes as particles linked together on chromosomes, structures found

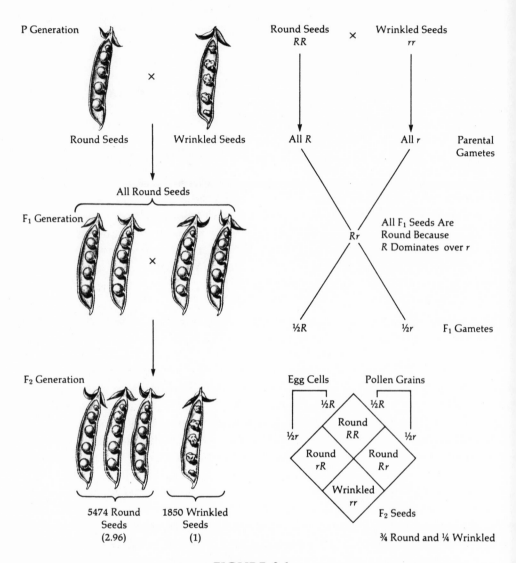

FIGURE 3.1

The Inheritence of Dominant and Recessive Traits in Mendel's Pea Plants

NOTE: Francisco Ayala and John Kiger, Jr., *Modern Genetics*, 2d ed. (Menlo Park, Calif.: Benjamin/Cummings, 1980), 34. Copyright © 1984 and 1980 by The Benjamin Cummings Publishing Company, Inc. Reprinted by permission.

inside the cell's nucleus: the most frequent image was of beads on a string. Continued experimentation, however, chipped away at this picture; great controversies erupted over the definition of a gene. Was it a rule for predicting phenotypic frequencies or a tangible physical entity?[6] Dr. G. Ledyard Stebbins, a well-known evolutionary biologist, points out that the

> ... entire history of Mendelian genetics has consisted of discoveries that have modified ... the Mendelian laws of heredity [which] have been so strongly modified by successive discoveries that in their original form they are useful only for elementary teaching.[7]

And herein lies one problem: because of these modifications, the word *gene* has several meanings.[8] Due to the confusion the word imparts to discussions of genetic causation, Stebbins provides a working definition:

> Biological heredity is transmitted by means of molecular templates, consisting of nucleic acids, that replicate with the aid of enzymes. These templates provide the code ... for the synthesis of specific proteins, the largest of class of working molecules of the body.[9]

Most frequently, the large molecule called DNA (deoxyribonucleic acid) forms the genetic material. Chromosomes, structures visible inside the cell's nucleus, consist of DNA in combination with other large molecules. The DNA molecule itself is a long chain composed of variously ordered repetitions of four differently shaped links called *bases.* The links, which behave as bases when dissolved singly in water, are chemically connected to molecules of phosphoric acid, which actually form the backbone of the long, spiral nuclei and molecule. The phosphoric acid component gives the DNA molecule as a whole the chemical properties of an acid. The order in which the links hang together forms a code, or as Stebbins calls it, a "molecular template." As part of the process of cell division, the DNA template, with the aid of enzymes, stamps out another copy of itself, so that when one cell divides into two each new cell contains the same DNA code as its parent. Biologists call this process *replication.*

Cells engage not only in replication, however, but also in protein synthesis; like DNA, proteins are also chainlike molecules. But instead of base links, the protein chain consists of connected amino acids. Not all of the more than twenty naturally occurring amino acids show up in every protein molecule, and many are repeated more than once in the same molecule. Thus each protein uniquely combines certain amino acids in a particular order on the protein chain. Two proteins consisting of the same kinds of amino acid may have very different chemical properties if the quantities of each type differ or if the amino acid links are arranged in different orders. The number and type of amino acids can combine in an almost infinite variety of different linear arrays, each a protein with special chemical properties, playing individual biological roles within an organism.

Triplets of bases linked together in particular orders on the DNA chain contain a code for particular amino acids. Imagine that the four bases found in DNA had the shapes of either a circle, a triangle, a square, or a trapezoid. Suppose that the order Circle-Square-Triangle provided the template for the synthesis of a particular amino acid, say, glutamic acid (O–□–△ = glutamic acid), while the order Square-Circle-Triangle) coded for a different amino acid, perhaps aspartic acid (□–O–△ = aspartic acid). A stretch of DNA that ran Circle-Square-Triangle-Square-Circle-Triangle (O–□–△–□–O–△) would code, then, for two links in the protein chain—glutamic acid followed by aspartic acid. During the past twenty-five years it has become clear that the four different links on the DNA chain arrange themselves in enough different triplet combinations to code for the twenty or so amino acids that combine in different numbers and orders to make up the thousands of different proteins in our bodies. Figures 3.2 and 3.3 provide illustrations of the DNA template and its role in DNA and protein synthesis. One can see from the drawings that the expression of the genetic code is complex, involving first the *transcription* of the code into an intermediate, "messenger" RNA molecule, followed by the *translation* of the transcribed code into a particular protein. Several different types of RNA (ribonucleic acids) as well as enzymes specialized for involvement in protein synthesis participate in these processes.*

All of this, though, is a far and abstract cry from Tiger's discussion of "the effect of genes on the situation of women," or even from the consideration of wrinkled peas or towheads. In order to connect our understanding of the genetic code and protein synthesis with our original questions about genes and behavior, we must move from the submicroscopic DNA molecules to visible,

---

* The drawing in figure 3.3 details the process of protein synthesis. Note that in addition to messenger RNA, *transfer RNA* (tRNA) molecules and *ribosome* structures consisting of *ribosomal* RNA and protein are also involved in the process.

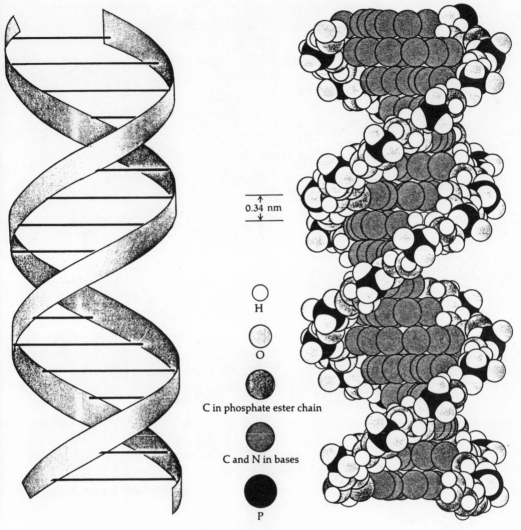

0.34 nm

H

O

C in phosphate ester chain

C and N in bases

P

FIGURE 3.2
The Structure of DNA
The DNA double helix: the ribbons of the model on the left and the strings of dark and shaded atoms in the space-filling model on the right represent the sugar-phosphate "backbones" of the two strands. The bases are stacked in the center of the molecule between the two backbones.

NOTE: Salvador E. Luria, Stephen Jay Gould, and Sam Singer, *A View of Life* (New York: Benjamin/Cummings, 1981), 43. Copyright © by The Benjamin/Cummings Publishing Company, Inc. Reprinted by permission.

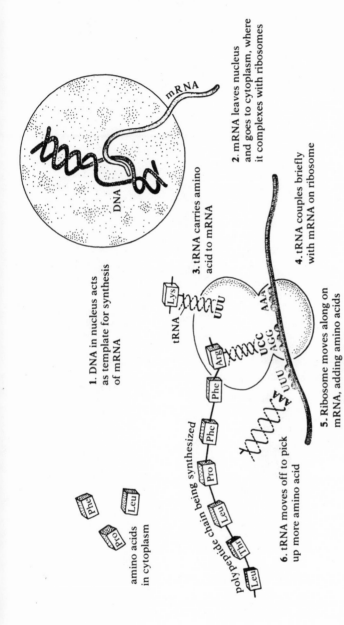

1. DNA in nucleus acts as template for synthesis of mRNA

2. mRNA leaves nucleus and goes to cytoplasm, where it complexes with ribosomes

3. tRNA carries amino acid to mRNA

4. tRNA couples briefly with mRNA on ribosome

5. Ribosome moves along on mRNA, adding amino acids to polypeptide chain

6. tRNA moves off to pick up more amino acid

amino acids in cytoplasm

polypeptide chain being synthesized

FIGURE 3.3

Protein Synthesis

Messenger RNA is synthesized, one chain of the DNA gene serving as the template. This mRNA then goes into the cytoplasm and becomes associated with ribosomes. The various types of tRNA in the cytoplasm pick up the amino acids for which they are specific and bring them to a ribosome as it moves along the mRNA. (The letters A, C, G, and U each stand for a particular base.) Each tRNA bonds to the mRNA at a point where a triplet of bases complementary to an exposed triplet on the tRNA occurs. This ordering of the tRNA molecules automatically orders the amino acids, which are then linked by peptide bonds. Synthesis of the polypeptide chain thus proceeds one amino acid at a time in an orderly sequence as the ribosomes move along the mRNA. As each tRNA donates its amino acid to the growing polypeptide chain, it uncouples from the mRNA and moves away into the cytoplasm, where it can be used again. (Note that the various molecules and organelles shown here are not drawn to scale with respect to one another or to the cell.)

NOTE: William T. Keeton, *Biological Science*, 3d ed., illus. Paula di Santo Bensadoun (New York: Norton, 1980), 647. Reproduced by permission of W. W. Norton & Company, Inc. Copyright © 1980, 1979, 1978, 1972, 1967 by W. W. Norton & Company, Inc.

easily observable traits. To illustrate, I will use an example from medical genetics in which the steps leading from DNA and protein to unmistakable clinical symptoms are well explained.

Consider the following case history. A patient complains of severe physical weakness. Her episodes of extreme pain seem worse when she exerts herself or travels to high altitudes; her anemic blood contains abnormal red cells which appear, under the microscope, to be bent into the shape of a half-moon or sickle. Even with proper care the patient may, over the years, experience increasingly severe attacks of weakness, at times showing impaired mental function, heart and kidney failure, and paralysis.

All of these symptoms characterize a genetically inherited disease called sickle-cell anemia. In adults a normally functioning red blood cell contains large amounts of the protein hemoglobin (called hemoglobin A, for Adult), which binds the oxygen taken into the lungs. This hemoglobin-bound oxygen, carried throughout the body by the circulating blood, keeps our cells breathing. The cells of a person afflicted with sickle-cell disease also produce hemoglobin, but in an abnormal form called hemoglobin S, which binds oxygen less efficiently than does the normal molecule. When more than enough oxygen is available, the presence of hemoglobin S may not matter, but under stressful conditions such as physical exertion or high altitude, hemoglobin S literally crystallizes inside the red blood cell. As a result of this change in the physical state and shape of the molecule, the entire red blood cell deforms to take on its characteristic sickle shape. The sickling itself has a variety of effects: the body's own defenses rapidly destroy the injured cells, causing severe anemia which can lead to physical weakness, impaired mental function, and heart failure. The misshapen cells may also get stuck in the narrow capillary blood vessels, causing severe pain as well as heart problems, brain, lung, and kidney damage, and possible paralysis (see figure 3.4).

Understanding the differences in physical properties of the normal and abnormal forms of hemoglobin gives us a good basis for comprehending how the disease generates its symptoms. But we understand even more because we know exactly what has happened to the DNA of a person with sickle-cell disease. Each of us carries two similar chromosomes (homologues) bearing a stretch of DNA with appropriate base sequences ready for transcription and translation into the sequence of amino acids that make up hemoglobin A. A person with sickle-cell disease, however, carries two homologues

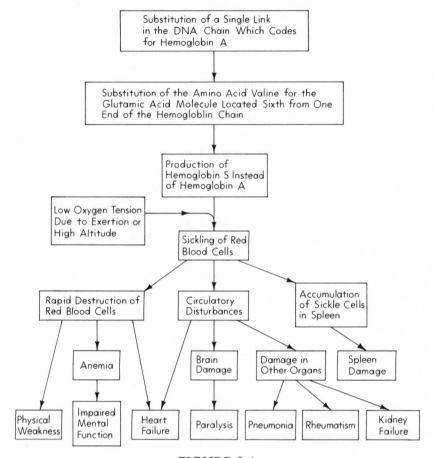

FIGURE 3.4

Chain of Events in Sickle-Cell Anemia

NOTE: The frequency of individual symptoms depends on chance environmental changes, quality of medical care, and individual differences.

in which one of the bases in the DNA sequences coding for hemoglobin has changed, or *mutated.* As a result, the sixth amino acid from one end of the hemoglobin molecule is transformed from glutamic acid to valine. In sum, this single base change in DNA results in a single amino acid substitution in a large protein molecule. The structure of the new amino acid changes the overall physical properties of the protein, so that it becomes less able to bind oxygen and less soluble inside the cell.

This example contains within it two important points. First, a simple alteration in the base sequence of DNA can have complex results. Figure 3.4 illustrates the cascade of events stemming from the single base change in question. The second point, which is perhaps the more important for our consideration of genes and behavior, is that even knowing that a person carries a specifically understood genetic defect does not enable us to predict the course of the disease in any particular individual. How much we can say with certainty depends entirely on the level of biological organization we choose to describe. If we describe the disease at the protein level, for example, we can safely state that all people who are homozygous for the genetic information coding for hemoglobin S will carry the hemoglobin S molecule: 100 percent predictability— a scientist's dream. But if we were to try to predict whether a particular sickle-cell patient would have leg ulcers (15 percent of them do), an enlarged heart (68 percent do), pneumonococcal meningitis (2 percent do), and so on, we would be at a loss for an answer.[10] At the level of the full-blown sickle-cell disease, each patient differs. The reasons are many, including differences among individuals in access to quantity and quality of medical care, different physical environments, different patterns of physical activity, different personality patterns, and individual differences in other (nonhemoglobin) genetic information that could directly or indirectly influence the course of the disease.

In short, while it is accurate shorthand to say that the sickle-cell mutation causes the production of an abnormal protein, it is wrong to say that it will cause any particular condition, for example, an enlarged heart, in any particular individual. At the whole, adult level of organization, genes alone do not produce biological phenotypes. Instead an individual's developmental and environmental history in combination with his or her total genetic endowment (all the genetic information encoded in the DNA),* as well as chance, contribute to the final phenotype. By the same token, genes alone do not determine human behavior. They work in the presence and under the influence of a set of environments.

---

* A staggering amount of information goes into the development of a human being. We all have twenty-three pairs of chromosomes (one set contributed from each parent) and each chromosome contains one enormously long DNA molecule. Estimates suggest that there is enough DNA in each of our cells to contain information for from ten thousand to fifty thousand different proteins.

Not By Genes Alone: The Brain and Behavior

Genes are thought to affect everything from our emotions to our legal system.[11] Researchers have asked whether males have a math gene, whether the Y chromosome causes aggressive behavior, and whether women are genetically programmed to have strong maternal feelings, to be verbally expressive, and to be better at relational than analytic skills. Harvard's E. O. Wilson suggests that "the genetic bias is intense enough to cause a substantial division of labor even in the most free and most egalitarian of future societies . . . even with identical education and equal access to all professions, men are likely to continue to play a disproportionate role in political life, business and science."[12] Genes apparently direct brain development which affects our behavior which determines our social structure. Although the reasoning is simple, little credible evidence actually exists to sustain such a chain of argument. Its logic, in truth, flies in the face of much of what we know about the biology of development.

If the brain codes for behavior and genes code for the brain, it must be a very complex cryptograph indeed, one for which biologists have yet to identify all of the basic symbols. The intricacy of the problem defies imagination. A human brain contains about 10^{10} (100 billion) cells, most of which have thousands of connections leading to or from other cells. Furthermore, each cell is itself chemically complex. One scientist estimates that during its lifetime every cell in the human brain must contend with the properties or activities of 10^9 giant molecules, including proteins, polysaccharides, and nucleic acids such as DNA.[13]

We do, nevertheless, have a degree of insight into brain development. Figure 3.5 provides an overview of some of its major known events. This picture is quite different from that describing sickle-cell anemia (see figure 3.4), in which the arrows connect a primary event—a change in base sequence in the DNA—to the appearance of a specific disease. In contrast to the relatively straightforward development of the red blood cell, the development of brain tissue is bewilderingly complex. Few of the biochemical details

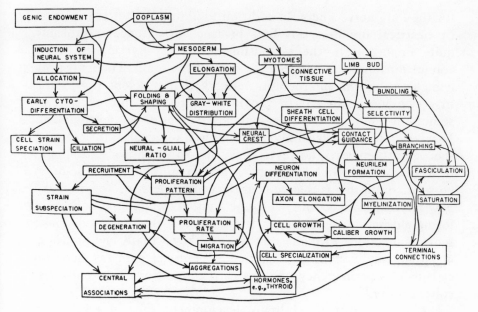

FIGURE 3.5

Causal Relations in Neurogenesis

NOTE: Benjamin H. Willier, Paul A. Weiss, and Victor Hamburger, *Analysis of Development* (Philadelphia: W. B. Saunders, 1955), 392.

of the events described in figure 3.4 are known, and one must keep in mind that each arrow in the drawing in reality runs through hundreds of boxes still unlabeled out of ignorance. Evidently we cannot envision brain development in terms of linear or even branching events. Instead we must think of a fine network, far more intricate than the random overlaying of hundreds of differently shaped spiderwebs.

Chastened by the level of the brain's complexity and the degree of our ignorance, let's look anyway at a few salient bits of knowledge. The human brain develops continuously in utero and continues the process of maturation for quite a number of years after birth. The total *number* of nerve cells is established during the first half of pregnancy.* But not so with all their interconnections, which continue to multiply during the first four years of life.[15] Glial cells, which continue to multiply in the brain and spinal cord for several years after birth, play an important role in making the so-called white matter, or myelin, material that provides electrical insulation

* Recent studies on adult birds suggest that nerve cells in adults may also increase in number.[14]

for the long nerve fibers. In addition to aiding in the transmission of electrical impulses, myelin's presence or absence is probably quite important in the establishment of the myriad cell-to-cell connections built up between the cells of the brain. The increasing complexity of the interneuron network that develops during the first two years after birth provides impressive confirmation of something obvious to any casual observer of child development. Newborn infants have only the simplest physical skills. During their first three years, however, they learn to sit up, walk, run, talk, feed themselves, climb jungle gyms, and so on. The inside-the-brain correlate of all this is that cells of the central nervous system make new physical connections with one another, while simultaneously losing old ones. One theory in fact holds that the brain cells make more connections with one another than they actually end up using, and that employment of a particular neuronal pathway helps to establish its permanency, the less used connections eventually dissolving away.[16]

A central feature of brain development thus emerges. *The physical structure of the adult brain—its size, number of cells, and most importantly its neuronal pathways—establishes itself in intimate interaction with the environment of the developing individual.* Nutrition, exercise, physical contact with other humans, exposure to varying sorts of visual and cognitive stimuli, all these and more influence brain structure. As Dr. T. Wiesel stated in his Nobel Prize lecture, "such sensitivity of the nervous system to the effects of experience may represent the fundamental mechanism by which the organism adapts to its environment during the period of growth and development."[17]

In addition to Wiesel's extensive work showing the role of postnatal experience in the development of the visual system, numerous other examples (some well defined, others merely suggestive) demonstrate the interdependence of body structure and environment. For instance, studies on the children of women who were pregnant during the Dutch famine of 1945 suggest that maternal starvation during the last third of gestation may slow the development of fat cells. The children born to these mothers were more likely than others to develop into thin adults and had fewer than average fat storage cells in their body. Furthermore, starvation during the first and second trimesters may affect the part of the brain, the hypothalamus, involved in regulating appetite. Children born of mothers starved during the first six months of pregnancy show a higher than average frequency of obesity because of over-eating.[18] Hence a casual observation that "obesity runs in families" tells us little about genetic causes, for the caloric intake of the mother—an environmental rather than a genetic factor—could influence both the structure and the number of fat cells as well as the eating behavior of the child.

There are other examples. One-day-old infants show exact, prolonged segments of movement correlating precisely in time with the speech patterns and rhythm of the adults talking to them.[19] This type of observation could go a long way toward explaining the transmission of physical mannerisms characteristic of individual families or even speech patterns of entire cultures.[20] Consider the studies showing that infants in certain West Kenyan farming communities learned to sit, stand, and walk at an earlier age than did American children to whom they were compared; observations revealed that the African mothers deliberately, through the use of play, forms of swaddling and carrying, and directly imposed practice, taught their babies each of these skills, in contrast to their American counterparts. The conclusion: we can properly consider the development of motor skills—often thought of by developmental psychologists as genetically directed—only in the context of a child's physical and social environment.[21]

A further complexity emerges from these examples, all three of which illustrate the ways in which a particular body structure develops only in a particular context. If the first take-home lesson in thinking about complex human traits is that linear chainlike causal explanations (genes code for the number of fat cells which in turn determine an individual's tendency toward obesity) are simply wrong, then the second is that the alternative idea of "environmental determinism" is also an oversimplification. My examples provide insight into the fact that the word *environment* itself has multiple meanings and numerous levels of complexity. In the example of the mothers in the Dutch famine, I suggested that one needs to look in utero to identify some of the critically relevant components of the environment; in the example of the babies mimicking adult movements, "environment" appears in the form of non-conscious adult behavior patterns critically transmitted during early infancy; while in the third example, the West Kenyan versus American mothers, "environment" turns up as a conscious teaching process. In all three cases the environmental factors described have done far more than establish some part of a psychological predisposition. They have interacted with a growing body to change the numbers and types of cells within it as well as to alter their connections to one another.

Modern psychology books often list four different theories of gender development—biological determinist, psychoanalytic, social learning, and cognitive developmental—writing about them as if they were mutually exclusive.[22] Examples of biological/environmental interactions teach us that the four theories are really like nonparallel lines: at some point, either close by or distant, they must intersect with one another. By trying to simplify and limit the intricacy of thought about human behavioral development, we trap ourselves into thinking at only one level or another, solely at the level of biology or of social learning.

Environment is not only multi-tiered, it is also without time limits. It exerts, rather, a cumulative continuum of effects over an entire lifetime. Studies in the United States and abroad, for example, have shown that severe malnutrition in infancy can cause mental retardation in later life. When one recalls the extensive brain development occurring during early childhood, such a finding makes some sense. It turns out, however, that when a child experiencing severe infant malnutrition subsequently grows up in an enriched social environment, the potential damage to IQ does not appear. Only severely malnourished children who *in addition* remain in economically and socially impoverished situations suffer from mental retardation.[23] In other words, one cannot surmise the effects of a condition encountered at one point in development without knowledge of the situation at other times in an individual's life. Each of us has many histories: the genetic histories of our ancestors; the chance encounter of a particular egg bearing some assortment of genetic information with a particular sperm bearing some other set; the racial/social, economic, and psychological histories of our families; our sex, birth order, and role within the family; our interactions with myriad adults, children, places, and schools as we grow up. In short all the chance events, from an inspiring concert attended at the last minute to a sudden death or severe accident or illness, form part of our individual histories which cumulatively change with each passing year. It is the sum of these events that becomes part of what for convenience's sake we call environment.

It may seem that we have digressed from our discussion of male and female development. The sexes are, after all, genetically different; the environment cannot change the presence of a Y or a pair of X chromosomes. But before returning to that issue, let us sum up. We have seen that genes are a historical concept supplanted in the era of molecular biology by knowledge of genetic information encoded in DNA. In a functioning cell the information may translate into a protein with the help of other molecules and multimolecular structures. If we define a trait or phenotype at the level of the translated protein, and if we consider all those genes involved in regulating the rates of synthesis and breakdown of both the specific protein in question and proteins in general, we can roughly state that the genotype determines the phenotype. If instead we consider more complex traits, occurring at supracellular levels of biological organization, the relatively simple correlation between genotype and phenotype breaks down. Two geneticists most graphically summarized the point when they wrote, "the hopelessness of the task of understanding behavior from single analytic approaches can be compared to the hopelessness of seeking linguistic insights by a chemical analysis of a book!"[24] Finally, a proper understanding of brain development suggests that while genetic information plays a key role in the unfolding of many details of the brain's structure, extensive development of nervous connections occurs after birth, influenced profoundly by individual experience. The brain's final "wiring diagram" as so many writers like to call it, is not a printed circuit, stamped out in accordance with a great genetic blueprint. It resembles more the weaving of an untutored artisan who, starting out with a general plan in mind, modifies it in the course of his or her work, using the available material and dyes, while covering up or making creative use of mistakes in pattern.

Nature Constructs Gender: X's, Y's, and the Hormones of Sexual Development

It is commonly held that sex is determined at the moment of conception. Eggs fertilized by Y-chromosome-bearing sperm develop into boys, those by X-chromosome-bearing sperm into girls. Behind the nursery rhyme's sugar, spice, snails, and puppy-dog tails stand male and female genes. Or so one hears. In truth, the roles in sexual development of the X and Y chromosomes are poorly understood.

Even those fetal events about which considerable information exists are described differently depending upon who writes the account; the story of the development in utero of male and female babies, fascinating in its own right, illustrates how deep cultural assumptions can condition the descriptive language used by scientists and in the process set limits on the scope of experimental exploration.

Consider the following real-life incidents:

1. An apparently normal boy baby has convulsions a few weeks after birth. Medical workups show that he has a disease that interferes with the adrenal glands' synthesis of the hormone cortisone. Further examination reveals that this "boy" has two X chromosomes, ovaries, oviducts, and a uterus. In fact "he" is really a girl whose vaginal lips have fused to become a scrotum and whose clitoris has developed into a penis. Genital surgery, treatment for the adrenal malfunction, and a name change produce a child who eventually grows up, marries, and has children.[25]

2. A child born in a small village in the West Indies works at her mother's side learning the female roles of washing and cooking, only to find at puberty that her voice deepens, a beard grows, and her clitoris enlarges into a penis. Eventually he takes a wife and fathers children.[26]

Both cases exemplify the unhinging of sexual development, the separation of gonadal, hormonal, and behavioral states from each other and from the X and Y chromosomes. When, where, and how do such changes originate? Studies of these situations furnish insights into the intricacies of sexual development while simultaneously fueling controversy over the biological bases of gender-related behavior.

Until the sixth week of development, XX and XY embryos are anatomically identical. During this early period each develops an embryonic gonad dubbed the "indifferent gonad" because of its sameness in both XX and XY embryos. In similar fashion the other internal structures of both the male and female reproductive systems begin to form, so that by the first month and a half of embryonic development all the embryos, regardless of which sex chromosomes reside inside their cells, have a set of female (Müllerian) ducts as well as a set of male (Wolffian) ducts, see figure 3.6. In XX embryos the female ducts normally develop into the oviducts, uterus, cervix,

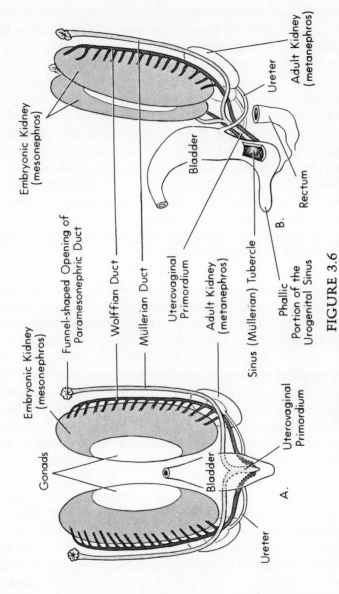

Embryonic Kidney
(mesonephros)

Embryonic Kidney
(mesonephros)

Gonads

Funnel-shaped Opening of
Paramesonephric Duct

Wolffian Duct

Müllerian Duct

Uterovaginal
Primordium

Uterovaginal
Primordium

Adult Kidney
(metanephros)

Bladder

Ureter

Bladder

Sinus (Müllerian) Tubercle

Phallic
Portion of the
Urogenital Sinus

Ureter

Adult Kidney
(metanephros)

Rectum

A.

B.

FIGURE 3.6

Internal Genitalia at the Indifferent Stage

NOTE: Keith L. Moore, *The Developing Human: Clinically Oriented Embryology* (Philadelphia: W. B. Saunders, 1973), 209.
A. is a sketch of a frontal view of the posterior abdominal wall of a seven-week embryo showing the two pairs of genital
ducts present during the indifferent stage. *B.* is a lateral view of a nine-week fetus showing the sinus (Müllerian) tubercle on
the posterior wall of the urogenital sinus.

and upper vagina, while in XY embryos the male ducts usually become the vas deferens, epididymis, and ejaculatory ducts. In other words, six weeks after fertilization all embryos are sexually bipotential, containing those parts needed to become either a male or a female. Although we have no idea which genes are involved in the process of reaching this early stage of sexual development, logically they must be present in both sexes.

So far, so good; now enter the Y chromosome. During the sixth week genetic information present on the Y becomes actively involved in promoting the synthesis of a protein called the H-Y antigen. Although the exact function of the antigen remains the subject of active research, it appears in humans to play some role in helping the tissues of the indifferent gonad to organize themselves into an embryonic testis.[27] The fetal testis contains recognizable sperm-producing tubules as well as the ability to synthesize hormones, two of which—testosterone and Müllerian Inhibiting Substance (MIS)—promote further events in the process of what has at this point clearly become male development.

Fetally synthesized testosterone, one of a group of chemically interrelated hormones called *androgens*, influences the male duct system to enter a period of active growth, while MIS, as its name suggests, ensures the degeneration of the still-present female duct system. Androgens and the chemically similar *estrogens* are not proteins but complex derivatives of the molecule cholesterol. The cell synthesizes them from a cholesterol base in an ordered series of steps, each governed by a different protein. Thus one or more sets of genes, coding for the proteins that catalyze each synthetic step, must be involved in the synthesis of one steroid hormone. These genetic messages are not located on the Y chromosome, being scattered instead on the X and one or more of the twenty-three pairs of non-sex chromosomes. In fact, as many as nineteen different genetic loci, not only on the X but on many different chromosomes, are known to be involved in the process of human sexual differentiation.[28] Although in males the Y must somehow involve itself with the selective translation of some of this information, individuals of both sexes remain genetically bipotential with regard to their ability to make androgens and estrogens. Stated in other terms, males and females have identical road maps for sexual development, but the Y chromosome helps the six-week-old male embryo choose the first of many branches taken in pursuing the destination of male development.

By comparison with all we know about male development, the genetic basis of female differentiation remains obscure. The most commonly told tale is that the fetal ovary produces no special sex hormones but that any embryo lacking testosterone develops in a female direction. Under usual conditions an XX gonad begins by the twelfth week following fertilization to develop into an ovary. The male duct system degenerates while the female system differentiates into the internal organs of the reproductive system. The view that females develop from mammalian embryos deficient in male hormone seems, oddly enough, to have satisfied the scientific curiosity of embryologists for some time, for it is the only account one finds in even the most up-to-date texts (see figure 3.7). Yet cracks have started to appear in this unanimous facade.[29] A recent review article, for instance, points out that the XX gonad begins synthesizing large quantities of estrogen at about the same time that the XY gonad begins to make testosterone. Just what does all that estrogen do? The authors report:

> Embryogenesis normally takes place in a sea of hormones . . . derived from the placenta, the maternal circulation, the fetal adrenal glands, the fetal testis, and possibly from the fetal ovary itself. . . . *It is possible that ovarian hormones are involved in the growth and maturation of the internal genitalia of the female.*[30] [Emphasis added]

Clearly, in contrast to the body of work done on male development, the final word on the genetic control of female development has yet to be written. That it has not yet been fully researched is due both to technical difficulties[31] and to the willingness of researchers to accept at face value the idea of passive female development. Because it is in the nature of research in developmental biology to always look for underlying causes, failure to probe beyond the "testosterone equals male"–"absence of testosterone equals female" hypothesis is a lapse which is at first difficult to understand. If, however, one notes the pervasiveness throughout all layers of our culture of the notion of "female as lack,"[32] then one learns from this account that such rock-bottom cultural ideas can intrude unnoticed even into the scientist's laboratory.

So far I've offered an account of the embryonic growth of the internal parts of the reproductive system. During an overlapping time period, however, the external genitalia also begin development. Until the eighth week of fetal life the external genitalia of the human embryo remain identical in both sexes. Figure 3.8 describes

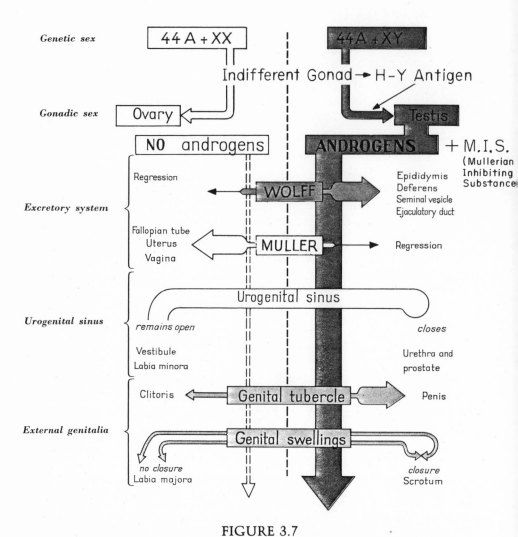

FIGURE 3.7

Textbook Accounts of Male and Female Embryonic Development

NOTE: H. Tuchmann-Duplessis et al., *Illustrated Human Embryology*, vol. 2, trans. L. S. Hurley (New York: Springer-Verlag, 1972), 100.

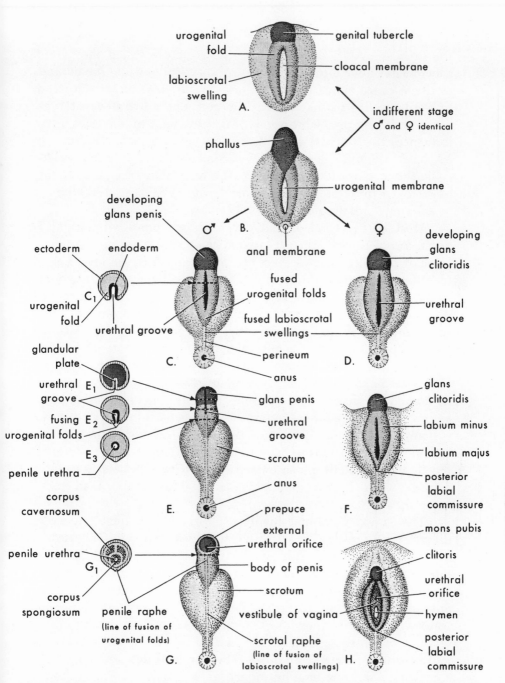

FIGURE 3.8

Development of the External Genitalia from the Indifferent Stage

NOTE: Keith L. Moore, *The Developing Human*, 2d ed. (Philadelphia: W. B. Saunders, 1977), 219. Diagrams *A* and *B* illustrate the development of the external genitalia during the indifferent stage (four to seven weeks). *C*, *E*, and *G* show the stages in the development of male external genitalia at about nine, eleven, and twelve weeks, respectively. To the left are schematic transverse sections (C_1, E_1 to E_3, and G_1) through the developing penis illustrating the formation of the penile urethra. *D*, *F*, and *H* show stages in the development of female external genitalia at nine, eleven, and twelve weeks, respectively.

the process by which the bipotential structure called the genital tubercle develops into either a penis or a clitoris, while the mounds of tissue called labioscrotal swellings develop into either a scrotum or the large lips of the vagina. The hormone dihydrotestosterone, secreted during the eighth week of development by the fetal testis, influences the external genitalia to develop in a male direction. In XX fetuses the female structures become evident by twelve weeks. As in the case of the internal reproductive system, the possible role of estrogen, progesterone, or other hormones in the development of external genitalia remains largely unexplored.

Most of the time, sexual development comes off without a hitch. Yet we began this discussion with two examples in which some of the different processes became unglued. In the first, a child identified as a boy on the basis of its external genitalia turned out to have two X chromosomes, ovaries, and oviducts. The cause of the confusion: a genetically inherited disease called adrenogenital syndrome (AGS). Individuals suffering from this illness lack an enzyme (protein) that normally converts precursor molecules (called 17-hydroxy-progesterone) into cortisone.[33] As a result large quantities of this precursor, not normally present in an XX embryo because of its immediate change into cortisone, become converted instead into androgens. This happens *after* the degeneration of the internal system of male ducts, but *before* the final formation of the external sex organs. The resulting baby bears two X chromosomes, internal female organs, a clitoris so enlarged that it may be mistaken for a penis (*is* a penis?), and labial folds that have fused to form an empty scrotum.

The converse example involves XY individuals unable to make dihydrotestosterone (DHT) during fetal development. This results in the birth of a child with female or ambiguous external genitalia, due to the missing DHT, but with male internal genitalia, because the synthesis of testosterone itself and of Müllerian Inhibiting Substance occurs normally (see figure 3.7). At puberty, however, testosterone rather than DHT serves as the masculinizing agent, and induces in these children the belated development of the penis and scrotum. These individuals have male genes and gonads, a somewhat abnormal fetal hormonal sex, a female infant anatomical sex, and a male adult hormonal and anatomical sex!

Clearly our binary, either/or categories of male and female do not work when we look at these genetic defects. Not even the sex organs are categorical. At what point in its growth do we stop calling the genital tubercle a clitoris and start calling it a penis? How small does a penis have to be before we call it a clitoris? And how many more experiments must we undertake before people stop writing that "the Adam Principle . . . governs the differentiation of the genital anatomy, namely that to differentiate a male something must be added"?[34] In summarizing an account of male and female development one might just as easily highlight their similarities, their identical starting points and minor divergences, and even sing the praises of some imagined underlying Female Principle that governs all development unless interfered with by some external agent such as testosterone. As with so many of the issues discussed herein, the facts are there for all to see. Which ones are emphasized and how the tale is told, however, seems to depend not on some objective truth but on the attitudes, be they conscious or subconscious, of the raconteur.

The Biology of Gender Revisited

In 1981 "The Sexes" made the cover of *Newsweek*. The definitive-sounding subtitle, "How They Differ and Why," heralded what turned out to be a far from conclusive article. Following an opening quotation (from playwright August Strindberg) alleging different biological natures for men and women, the authors wrote:

> Research on the structure of the brain, on the effects of hormones, and in animal behavior, child psychology and anthropology is providing new scientific underpinnings for what August Strindberg and his ilk viscerally guessed: men and women *are* different.

The article, though, concluded with the following:

> Perhaps the most arresting implication of the research up to now is not that there are undeniable differences between males and females but that their differences are so small relative to the possibilities open to them.[35]

The take-home message is hard to fathom. Are males and females different or aren't they? The authors of the *Newsweek* article hedged

85

their bets, and well they might, for reasons that lie deep in the nature of research on the biology of gender. Nothing could illustrate this more clearly than the current debate about the hormonal basis of gender identity.

At its heart the controversy centers on that hardy perennial, the nature-versus-nurture debate. On one side stand psychologist Dr. John Money and his associates, proponents of the idea that gender identity becomes fixed during the first three years of a child's life, depending primarily on the sex of rearing rather than on such details as the presence or absence of a Y chromosome, ovaries, or the like. On the opposite side, one finds physician and researcher Dr. Julianne Imperato-McGuinley and her co-workers. They assert that gender identity can remain flexible throughout childhood, but that the hormones of puberty fix it irrevocably. To them the sex of rearing may be less important than the sex hormones that appear during adolescence. To top it all off, each group has apparently compelling evidence to support its position.

Money has devoted a considerable portion of his career to examining and following up on children, such as those affected by adrenogenital syndrome, whose correct chromosomal sex was identified only months or years into childhood. He found that sex reassignment was easily effected if it occurred during the first three years of life, but that later changes were psychologically difficult if not impossible. A child raised as a boy until age twelve, for example, experienced deep distress when he developed breasts. Although the physical evidence of puberty suggested that he might be a girl, he was not swayed and opted to have a double mastectomy in order to clear up the confusion.[36] In general Money found that pairs of children matched for identity of physical development accepted opposite gender identities if they were raised as members of the opposite sex.

Imperato-McGuinley also studied sexually incongruous individuals, most of whom came from three rural villages in Santo Domingo. The people in her study had the above-mentioned genetically inherited deficiency for the androgen dihydrotestosterone. While all thirty-eight of those identified as DHT-deficient had ambiguous genitalia, eighteen were raised as females. At puberty, as their voices deepened, these "girls" grew adult-sized penises and scrotums. The phenomenon, which would certainly surprise most North Americans, is common enough in these little towns that the villagers have coined a word, literally translated as "penis-at-twelve," to describe such children. Of the eighteen raised as girls, sixteen seem to have successfully assumed a male gender role, marrying and fathering children.

The Dominican research team also paid attention to the social aspects of being male or female in a small Santo-Domingan village. The roles set aside for each sex are strict, with males permitted a great deal more freedom and social life outside the home. "The girls are encouraged to stay with their mothers or play near the house. . . . After 11 or 12 years of age the boys seek entertainment at the bars . . . the girls . . . stay home and help with the household chores. . . . Fidelity is demanded from the women but not from the men."[37] Thus one could imagine that having the option of becoming a male might have its attractions to a young teenager raised as a female. This would especially be true if that female had abnormal genitalia and knew of the penis-at-twelve syndrome. The degree of conscious articulation of difference—the awareness of belonging to a third sex—among the DHT-deficient children raised as girls or among their parents is unclear. Imperato-McGuinley simply says they were unambiguously raised as girls. Unless one does not believe in subconsciously motivated behavior, however, it is a bit difficult to accept the claim that these children were raised just like any genitally normal girl.

Imperato-McGuinley and her co-workers drew two conclusions, both apparently at variance with Money's. First, "gender identity is not unalterably fixed in childhood but is continually evolving," and second, "when the sex of rearing is contrary to the testosterone-mediated biologic sex, the biologic sex prevails."[38] Their first conclusion is accurate only if one assumes that neither the parents nor the children raised as girls realized that they had the penis-at-twelve syndrome. If they did know, it is possible that the children were raised with the option of future maleness built into their psychological development. If this were the case, then these children do not necessarily provide data in conflict with Money's. On the other hand, Money's notion of irrevocable psychological fixation may itself be too rigidly stated. Humans, especially children, are full of surprises. And so it may be with the development of gender identity.

Ferreting out the meaning of the second conclusion is a bit more difficult. Imperato-McGuinley seems to believe that hormones can override extensive early psychological programming. Examination of the data and a literal reading of her words, however, reveals a second meaning. She writes that when the option is to choose the *testosterone-mediated* sex—that is, when a child raised as a girl has a chance to become a boy—then it makes this choice. The reasons for that choice *could*, of course, be entirely hormonal, but given that in these villages boys are freer and that these children *look* like boys, one can easily imagine a nonbiological mediation of choice. Both meanings appear, inseparably, in the data. The preferred interpretation is up to us, but we have only our sociopolitical instincts on which to lean in exercising our option. The hard facts permit no final judgment, for a key bit of information is missing: Are there biologically mediated situations under which children raised as *boys* would choose to become *girls*, even though the social setting permits males greater freedom? Without such a natural control one cannot unscramble the implications of Imperato-McGuinley's data.

Sex, then, is no simple matter. Most often all the components of biologic sex concur and an infant, identified at birth as either male or female, is raised accordingly. Cases in which biologic and assigned sexes differ reveal enormous psychological flexibility. Some children, perhaps even to the age of puberty, can cope with changing their sex, provided at least that the change agrees with their anatomy. But it is difficult for young people to handle a disagreement between their external anatomical sex and their assigned sex—not a particularly surprising observation. In all of the cases studied both by Money's and by Imperato-McGuinley's groups, either the assigned sex changed spontaneously to agree with the visible anatomy or doctors surgically altered the anatomy to bring it more into line with the assigned sex. Whichever the direction of change, though, the human hallmark of enormous biological complexity and psychological flexibility shines through.

Conclusion: From Gene to Gender

We began this chapter by trying to figure out whether it makes any sense to talk about genes that control behavior. Let us conclude by stating that a "pure" biological explanation of anything as complex and unpredictable as human behavior would by its very nature be unequal to the task. Genetic information specifically encodes for the amino acid sequences of particular proteins. Complex traits, however, arise not simply by the accretion of a set of macromolecules; they form part of yet larger and more complex physical forms such as cells, tissues, and organs. To do so they must be present in the right amounts at the right time and be organized in the correct spatial configurations. All of this is subject to three influences: genetic regulatory information, intrusion from the external environment, and chance variations in development. In other words, referring to a genetic ability to perform math or music, or to a biological tendency toward aggressive behavior obscures rather than informs. To understand human development we need to know a great deal more about how the environment affects physical growth and patterns, and how individual variation including genetic variation plays into each different life history to produce adults with different competencies and potentials.

The discussion of sexual development illustrates two points: the first, reiterated throughout this book, is that a scientist's cultural vision can limit his or her scientific vision. What research has offered us to date under the rubric of "sexual development" is an analytical account of *male* development. The absence of the second sex seems to have gone unnoticed in theories of sexual differentiation. The second point is that children show a great deal of (albeit not total) flexibility in the development of a gender self-concept. Studies of fetal development, however, as well as analyses of the differing reproductive physiologies of adult males and females raise a series of related but as yet unresolved questions. Some scientists hold, for instance, that male/female differences in behavior stem from different hormonal environments encountered in utero. Indeed, in other animals embryonic exposure to testosterone sometimes changes the physical structure of the brain. But even if human behavior were *unaffected* by fetal androgens, surely the vast differences in hormone levels circulating in the bloodstreams of male and female adults could influence our behavior. Or so it is said. It is to these issues that we turn our attention in the next two chapters, looking first at possible behavioral changes in women linked to the phases of the menstrual cycle and menopause, and then examining the idea that testosterone, either through action in utero or in adults, makes men more aggressive than women.

4

HORMONAL HURRICANES: MENSTRUATION, MENOPAUSE, AND FEMALE BEHAVIOR

Woman is a pair of ovaries with a human being
attached, whereas man is a human being furnished
with a pair of testes.
—RUDOLF VIRCHOW, M.D. (1821–1902)

Estrogen is responsible for that strange mystical
phenomenon, the feminine state of mind.
—DAVID REUBEN, M.D., 1969

IN 1900, the president of the American Gynecological Association eloquently accounted for the female life cycle:

Many a young life is battered and forever crippled in the breakers of puberty; if it crosses these unharmed and is not dashed to pieces on the rock of childbirth, it may still ground on the ever-recurring shadows of menstruation and lastly upon the final bar of the menopause ere protection is found in the unruffled waters of the harbor beyond the reach of the sexual storms.[1]

Since then we have amassed an encyclopedia's worth of information about the existence of hormones, the function of menstruation, the regulation of ovulation, and the physiology of menopause. Yet

90

many people, scientists and nonscientists alike, still believe that women function at the beck and call of their hormonal physiology. In 1970, for example, Dr. Edgar Berman, the personal physician of former Vice President Hubert Humphrey, responded to a female member of congress:

> Even a Congresswoman must defer to scientific truths . . . there just are physical and psychological inhibitants that limit a female's potential. . . . I would still rather have a male John F. Kennedy make the Cuban missile crisis decisions than a female of the same age who could possibly be subject to the curious mental aberrations of that age group.[2]

In a more grandiose mode, Professor Steven Goldberg, a university sociologist, writes that "men and women differ in their hormonal systems . . . every society demonstrates patriarchy, male dominance and male attainment. The thesis put forth here is that the hormonal renders the social inevitable."[3]

At the broadest political level, writers such as Berman and Goldberg raise questions about the competency of *any and all* females to work successfully in positions of leadership, while for women working in other types of jobs, the question is, Should they receive less pay or more restricted job opportunities simply because they menstruate or experience menopause? And further, do women in the throes of premenstrual frenzy frequently try to commit suicide? Do they really suffer from a "diminished responsibility" that should exempt them from legal sanctions when they beat their children or murder their boyfriends?[4] Is the health of large numbers of women threatened by inappropriate and even ignorant medical attention—medical diagnoses that miss real health problems, while resulting instead in the prescription of dangerous medication destined to create future disease? These issues among others will be addressed in this chapter.

The idea that women's reproductive systems direct their lives is ancient. But whether it was Plato, writing about the disruption caused by barren uteri wandering about the body,[5] Pliny, writing that a look from a menstruating woman will "dim the brightness of mirrors, blunt the edge of steel and take away the polish from ivory,"[6] or modern scientists writing about the changing levels of estrogen and progesterone, certain messages emerge quite clearly. Women, by nature emotionally erratic, cannot be trusted in positions of responsibility. Their dangerous, unpredictable furies warrant control by the medical profession,* while ironically, the same "dangerous" females also need protection because their reproductive systems, so necessary for the procreation of the race, are vulnerable to stress and hard work.

"The breakers of puberty," in fact, played a key role in a debate about higher education for women, a controversy that began in the last quarter of the nineteenth century and still echoes today in the halls of academe. Scientists of the late 1800s argued on physiological grounds that women and men should receive different types of education. Women, they believed, could not survive intact the rigors of higher education. Their reasons were threefold: first, the education of young women might cause serious damage to their reproductive systems. Energy devoted to scholastic work would deprive the reproductive organs of the necessary "flow of power," presenting particular problems for pubescent women for whom the establishment of regular menstruation was of paramount importance. Physicians cited cases of women unable to bear children because they pursued a course of education designed for the more resilient young man.[7] In an interesting parallel to modern nature-nurture debates, proponents of higher education for women countered biological arguments with environmental ones. One anonymous author argued that, denied the privilege afforded their brothers of romping actively through the woods, women became fragile and nervous.[8]

Opponents of higher education for women also claimed that females were less intelligent than males, an assertion based partly on brain size itself but also on the overall size differences between men and women. They held that women cannot "consume so much food as men ... [because] their average size remains so much smaller; so that the sum total of food converted into thought by women can never equal the sum total of food converted to thought by men. It follows therefore, that *men will always think more than women.*"[9] One respondent to this bit of scientific reasoning asked the thinking reader to examine the data: Aristotle and Napoleon were short, Newton, Spinoza, Shakespeare, and Comte delicate and

---

* In the nineteenth century, control took the form of sexual surgery such as ovariectomies and hysterectomies, while twentieth-century medicine prefers the use of hormone pills. The science of the 1980s has a more sophisticated approach to human physiology, but its political motives of control and management have changed little. For an account of medicine's attitudes toward women, see Barbara Ehrenreich and Deidre English, *For Her Own Good: 150 Years of Experts' Advice to Women* (New York: Doubleday, 1979); and G. J. Barker-Benfield, *The Horrors of the Half-Known Life* (New York: Harper & Row, 1977).

of medium height, Descartes and Bacon sickly, "while unfortunately for a theory based upon superior digestion, Goethe and Carlyle were confirmed dyspeptics."[10] Finally, as if pubertal vulnerability and lower intelligence were not enough, it seemed to nineteenth-century scientists that menstruation rendered women "more or less sick and unfit for hard work" "for one quarter of each month during the best years of life."[11]

Although dated in some of the particulars, the turn-of-the-century scientific belief that women's reproductive functions make them unsuitable for higher education remains with us today. Some industries bar fertile women from certain positions because of workplace hazards that might cause birth defects, while simultaneously deeming equally vulnerable men fit for the job.* Some modern psychologists and biologists suggest that women perform more poorly than do men on mathematics tests because hormonal sex differences alter male and female brain structures; and many people believe women to be unfit for certain professions because they menstruate. Others argue that premenstrual changes cause schoolgirls to do poorly in their studies, to become slovenly and disobedient, and even to develop a "nymphomaniac urge [that] may be responsible for young girls running away from home . . . only to be found wandering in the park or following boys."[12]

If menstruation really casts such a dark shadow on women's lives, we ought certainly to know more about it—how it works, whether it can be controlled, and whether it indeed warrants the high level of concern expressed by some. Do women undergo emotional changes as they progress through the monthly ovulatory cycle? And if so, do hormonal fluctuations bring on these ups and downs? If not—if a model of biological causation is inappropriate—how else might we conceptualize what happens?

The Shadows of Menstruation: A Reader's Literature Guide

The Premenstrual Syndrome

SCIENCE UPDATE: PREMENSTRUAL STRAIN LINKED TO CRIME
—*Providence Journal*

* The prohibited work usually carries a higher wage.

ERRATIC FEMALE BEHAVIOR TIED TO PREMENSTRUAL SYNDROME
—*Providence Journal*
VIOLENCE BY WOMEN IS LINKED TO MENSTRUATION
—*National Enquirer*

Menstruation makes news, and the headlines summarize the message. According to Dr. Katharina Dalton (see chapter 1), Premenstrual Syndrome (PMS) is a medical problem of enormous dimensions. Under the influence of the tidal hormonal flow, women batter their children and husbands, miss work, commit crimes, attempt suicide, and suffer from up to 150 different symptoms, including headaches, epilepsy, dizziness, asthma, hoarseness, nausea, constipation, bloating, increased appetite, low blood sugar, joint and muscle pains, heart palpitations, skin disorders, breast tenderness, glaucoma, and conjunctivitis.[13] Although the great concern expressed in the newspaper headlines just quoted may come from a single public relations source,[14] members of the medical profession seem eager to accept at face value the idea that "70 to 90% of the female population will admit to recurrent premenstrual symptoms and that 20 to 40% report some degree of mental or physical incapacitation."[15]

If all this is true, then we have on our hands nothing less than an overwhelming public health problem, one that deserves a considerable investment of national resources in order to develop understanding and treatment. If, on the other hand, the claims about premenstrual tension are cut from whole cloth, then the consequences are equally serious. Are there women in need of proper medical treatment who do not receive it? Do some receive dangerous medication to treat nonexistent physiological problems? How often are women refused work, given lower salaries, taken less seriously because of beliefs about hormonally induced erratic behavior? In the game of PMS the stakes are high.

The key issues surrounding PMS are so complex and interrelated that it is hard to know where to begin. There is, as always, the question of evidence. To begin with we can look, in vain, for credible research that defines and analyzes PMS. Despite the publication of thousands of pages of allegedly scientific analyses, the most recent literature reviews simultaneously lament the lack of properly done studies and call for a consistent and acceptable research definition and methodology.[16] Intimately related to the question of evidence is that of conceptualization. Currently held theoretical views about the reproductive cycle are inadequate to the task of understanding the emotional ups and downs of people functioning in a complex world. Finally, lurking beneath all of the difficulties of research design, poor methods, and muddy thinking is the medical world's view of the naturally abnormal woman. Let's look at this last point first.

If you're a woman you can't win. Historically, females who complained to physicians about menstrual difficulties, pain during the menstrual flow, or physical or emotional changes associated with the premenstruum heard that they were neurotic. They imagined the pain and made up the tension because they recognized menstruation as a failure to become pregnant, to fulfill their true role as a woman.[17] With the advent of the women's health movement, however, women began to speak for themselves.[18] The pain is real, they said; our bodies change each month. The medical profession responded by finding biological/hormonal causes, proposing the need for doctor-supervised cures. A third voice, however, entered in: that of feminists worried about repercussions from the idea that women's natural functions represent a medical problem capable of preventing women from competing in the world outside the home. Although this multisided discussion continues, I currently operate on the premise that some women probably do require medical attention for incapacitating physical changes that occur in synchrony with their menstrual cycle. Yet in the absence of any reliable medical research into the problem it is impossible to diagnose true disease or to develop rational treatment. To start with, we must decide what is normal.

The tip-off to the medical viewpoint lies in its choice of language. What does it mean to say "70 to 90% of the female population will admit to recurrent premenstrual symptoms"?[19] The word *symptom* carries two rather different meanings. The first suggests a disease or an abnormality, a condition to be cured or rendered normal. Applying this connotation to a statistic suggesting 70 to 90 percent symptom formation leads one to conclude that the large majority of women are by their very nature diseased. The second meaning of *symptom* is that of a sign or signal. If the figure of 70 to 90 percent means nothing more than that most women recognize signs in their own bodies of an oncoming menstrual flow, the statistics are unremarkable. Consider then the following, written in 1974 by three scientists:

> It is estimated that from 25% to 100% of women experience some form of premenstrual or menstrual emotional disturbance. . . . Eichner makes the discerning point that the few women who do not admit to premenstrual tension are basically unaware of it but one only has to talk to their husbands or co-workers to confirm its existence.[20]

Is it possible that up to 100 percent of all menstruating women regularly experience emotional disturbance? Compared to whom? Are males the unstated standard of emotional stability? If there is but a single definition of what is normal and men fit that definition, then women with "female complaints" must by definition be either crazy or in need of medical attention. A double bind indeed.

Some scientists explicitly articulate the idea of the naturally abnormal female. Professor Frank Beach, a pioneer in the field of animal psychology and its relationship to sexuality, suggests the following evolutionary account of menstruation. In primitive hunter-gatherer societies adult women were either pregnant or lactating, and since life spans were so short they died well before menopause; low-fat diets made it likely that they did not ovulate every month; they thus experienced no more than ten menstrual cycles. Given current life expectancies as well as the widespread use of birth control, modern women may experience a total of four hundred menstrual cycles. He concludes from this reasoning that "civilization has given women *a physiologically abnormal status* which may have important implications for the interpretation of psychological responses to periodic fluctuations in the secretion of ovarian hormones"—that is, to menstruation (emphasis added).[21] Thus the first problem we face in evaluating the literature on the premenstrual syndrome is figuring out how to deal with the underlying assumption that women have "a physiologically abnormal status."

Researchers who believe in PMS hold a wide variety of viewpoints (none of them supported by scientific data) about the basis of the problem. For example, Dr. Katharina Dalton, the most militant promoter of PMS, says that it results from a relative end-of-the-cycle deficiency in the hormone progesterone. Others cite deficiencies in vitamin B-6, fluid retention, and low blood sugar as possible causes. Suggested treatments range from hormone injection to the use of lithium, diuretics, megadoses of vitamins, and control of sugar in the diet[22] (see table 4.1 for a complete list). Although some of these treatments are harmless, others are not. Progesterone injection causes cancer in animals. What will it do to humans? And a recent issue of *The New England Journal of Medicine* contains a

TABLE 4.1

Alleged Causes and Proposed Treatments of PMS

Hypothesized Causes of Premenstrual Syndrome	Various PMS Treatments (Used But Not Validated)
Estrogen excess	Oral contraceptives (combination estrogen
Progesterone deficiency	and progesterone pills)
Vitamin B deficiency	Estrogen alone
Vitamin A deficiency	Natural progesterone
Hypoglycemia	Synthetic progestins
Endogenous hormone allergy	Valium or other tranquilizers
Psychosomatic	Nutritional supplements
Fluid retention	Minerals
Dysfunction of the neurointermediate	Lithium
lobe of the pituitary	Diuretics
Prolactin metabolism	A prolactin inhibitor/dopamine agonist
	Exercise
	Psychotherapy, relaxation, education,
	reassurance

SOURCES: Robert L. Reid and S. S. Yen, "Premenstrual Syndrome," *American Journal of Obstetrics and Gynecology* 139(1981):85–104; and Judith Abplanalp, "Premenstrual Syndrome: A Selective Review," *Women and Health* 8(1983):107–24.

report that large doses of vitamin B-6 damage the nerves, causing a loss of feeling in one's fingers and toes.[23] The wide variety of PMS "causes" and "cures" offered by the experts is confusing to put it mildly. Just what *is* this syndrome that causes such controversy? How can a woman know if she has it?

With a case of the measles it's really quite simple. A fever and then spots serve as diagnostic signs. A woman said to have PMS, however, may or may not have any of a very large number of symptoms. Furthermore, PMS indicators such as headaches, depression, dizziness, loss or gain of appetite show up in everyone from time to time. Their mere presence cannot (as would measle spots) help one to diagnose the syndrome. In addition, whether any of these signals connote disease depends upon their severity. A slight headache may reflect nothing more than a lack of sleep, but repeated, severe headaches could indicate high blood pressure. As one researcher, Dr. Judith Abplanalp, succinctly puts it: "There is no one set of symptoms which is considered to be the hallmark of or standard criterion for defining the premenstrual syndrome."[24] Dr. Katharina Dalton agrees but feels one can diagnose PMS quite simply by applying the term to "any symptoms or complaints which regularly come just before or during early menstruation but are absent at other times of the cycle."[25] Dalton contrasts this with men suffering from potential PMS "symptoms," because, she says, they experience them randomly during the month while women with the same physical indications acknowledge them only during the premenstruum.

PMS research usually bases itself on an ideal, regular, twenty-eight-day menstrual cycle. Researchers eliminate as subjects for study women with infrequent, shorter, or longer cycles. As a result, published investigations look at a skewed segment of the overall population. Even for those women with a regular cycle, however, a methodological problem remains because few researchers define the premenstrual period in the same way. Some studies look only at the day or two preceding the menstrual flow, others look at the week preceding, while workers such as Dalton cite cases that begin two weeks before menstruation and continue for one week after. Since so few investigations use exactly the same definition, research publications on PMS are difficult to compare with one another.[26] On this score if no other, the literature offers little useful insight, extensive as it is.

Although rarely stated, the assumption is that there is but *one* PMS. Dalton defines the problem so broadly that she and others may well lump together several phenomena of very different origins, a possibility heightened by the fact that investigators rarely assess the severity of the symptoms. Two women, one suffering from a few low days and the other from suicidal depression, may both be diagnosed as having PMS. Yet their difficulties could easily have different origins and ought certainly to receive different treatments. When investigators try carefully to define PMS, the number of people qualifying for study decreases dramatically. In one case a group used ten criteria (listed in table 4.2) to define PMS only to find that no more than 20 percent of those who had volunteered for their research project met them.[27] In the absence of any clearly agreed upon definition(s) of PMS, examinations of the topic should at least state clearly the methodology used; this would enable comparison between publications, and allow us to begin to accumulate some knowledge about the issues at hand (table 4.2 lists suggested baseline information). At the moment the literature is filled with individual studies that permit neither replication nor comparison with one another—an appropriate state, perhaps, for an art gallery but not for a field with pretensions to the scientific.

Despite the problems of method and definition, the conviction remains that PMS constitutes a widespread disorder, a conviction

TABLE 4.2

Toward a Definition of Premenstrual Syndrome

Experimental Criteria (Rarely Met in PMS Studies)

Premenstrual symptoms for at least six preceding cycles

Moderate to severe physical and psychological symptoms

Symptoms *only* during the premenstrual period with marked relief at onset of menses

Age between 18 and 45 years

Not pregnant

No hormonal contraception

Regular menses for six previous cycles

No psychiatric disorder; normal physical examination and laboratory test profile

No drugs for preceding four weeks

Will not receive anxiolitics, diuretics, hormones, or neuroleptic drugs during the study

---

Minimal Descriptive Information to Be Offered in Published Studies of PMS
    (Rarely Offered in the Current Literature)

Specification of the ways in which subjects were recruited

Age limitations

Contraception and medication information

Marital status

Parity

Race

Menstrual history data

Assessment instruments

Operational definition of PMS

Psychiatric history data

Assessment of current psychological state

Criteria for assessment of severity of symptoms

Criteria for defining ovulatory status of cycle

Cut-off criteria for "unacceptable" subjects

SOURCE: Judith Abplanalp, "Premenstrual Syndrome: A Selective Review," *Women and Health* 8(1983):107–24.

that fortifies and is fortified by the idea that women's reproductive function, so different from that of "normal" men, places them in a naturally diseased state. For those who believe that 90 percent of all women suffer from a disease called PMS, it becomes a reasonable research strategy to look at the normally functioning menstrual cycle for clues about the cause and possible treatment. There are, in fact, many theories but no credible evidence about the origins of PMS. In table 4.1 I've listed the most frequently cited hypotheses, most of which involve in some manner the hormonal system that regulates menstruation. Some of the theories are ingenious and require a sophisticated knowledge of human physiology to comprehend. Nevertheless, the authors of one recent review quietly offer the following summary: "To date no one hypothesis has adequately explained the constellation of symptoms composing PMS."[28] In short, PMS is a disease in search of a definition and cause.

PMS also remains on the lookout for a treatment. That many have been tried is attested to in table 4.1. The problem is that only rarely has the efficacy of these treatments been tested with the commonly accepted standard of a large-scale, double-blind study that includes placebos. In the few properly done studies "there is usually (1) a high placebo response and (2) the active agent is usually no better than a placebo."[29] In other words, women under treatment for PMS respond just as well to sugar pills as to medication containing hormones or other drugs. Since it is probable that some women experience severe distress caused by malfunctions of their menstrual system, the genuinely concerned physician faces a dilemma. Should he or she offer treatment until the patient says she feels better even though the drug used may have dangerous side effects; or should a doctor refuse help for as long as we know of no scientifically validated treatment for the patient's symptoms? I have no satisfactory answer. But the crying need for some scientifically acceptable research on the subject stands out above all. If we continue to assume that menstruation is itself pathological, we cannot establish a baseline of health against which to define disease. If, instead, we accept in theory that a range of menstrual normality exists, we can then set about designing studies that define the healthy female reproductive cycle. Only when we have some feeling for *that* can we begin to help women who suffer from diseases of menstruation.

Many of those who reject the alarmist nature of the publicity surrounding PMS believe nevertheless that women undergo mood changes during their menstrual cycle. Indeed most Western women would agree. But do studies of large segments of our population support this generality? And if so, what causes these ups and downs? In trying to answer these questions we confront another piece of the medical model of human behavior, the belief that biology is primary, that hormonal changes cause behavioral ones, but not vice versa. Most researchers use such a linear, unicausal model without thinking about it. Their framework is so much a part of their belief system that they forget to question it. Nevertheless it is the model from which they work, and failure to recognize and work skeptically with it often results in poorly conceived research combined with implausible interpretations of data. Although the paradigm of biological causation has until very recently dominated menstrual cycle research, it now faces serious and intellectually stimulating challenge from feminist experts in the field.

Lifting the Shadow: Toward an Understanding of Menstruation and Behavior

During the forty-five years that have elapsed since Benedek and Rubenstein published the first "modern" experimental study of mood and menstrual cycle,[30] countless additional studies have appeared. Why, then, is it possible for a scientist in 1981 to write the following?

> The question is still being debated. Are women's moods a function of their rhythmic reproductive physiology or not? Unfortunately the literature does not agree on the answer to the more basic question: Do women experience cyclic mood fluctuations?[31]

In looking for answers, we encounter a research field filled with poorly designed studies. Inadequate sample sizes and measures, inappropriate choice of subjects, tests designed to obtain a desired outcome, and poor or nonexistent use of statistical analysis are but some of the problems. That so many scientists have been able for so long to do such poor research attests to both the unconscious social agendas of many of the researchers and to the theoretical inadequacy of the research framework used in the field as a whole. Once again we encounter the failure of a simple linear model of biological causation, and must struggle instead with a more complex conceptualization in which mind, body, and culture depend so inextricably on one another that allegedly straightforward studies, ones claiming to find single causes for cyclic behavior, must be looked upon with deep suspicion.

Over the years specific critiques of menstrual-cycle studies have accumulated. As early as 1877 physician and medical researcher Mary Putnam Jacobi pioneered menstrual-cycle research with the publication of her book *The Question of Rest for Women During Menstruation*. In this critical literature review she pointed out the relative novelty of the idea that women are *more* rather than *less* vulnerable at menstruation. In a thorough analysis of her own careful interviews with 268 women about their menstrual experiences, she presented results that have a ring of feminist modernity, suggesting that by no means do all women suffer menstrual pain and that among those who do are many who are otherwise in poor health or under stress, or who have insufficient exercise and education.[32]

Psychologist Mary Brown Parlee deserves the credit for breaking ground more recently with her analyses of the menstrual cycle. Her review article, "The Premenstrual Syndrome," published in 1973, systematically takes to task the research of the preceding thirty-five years and sets the stage for even more sophisticated theoretical and methodological analyses.[33] Parlee categorizes studies on premenstrual emotionality into four types: correlational, retrospective question-naires, daily self-reports, and thematic analysis of word lists describing various feelings and moods.

Correlational studies. Through correlational studies, scientists look to see if particular phases of the menstrual cycle coincide with reported or observed behaviors. Ideally investigators try to determine the phase of the menstrual cycle independently, using some reliable measure such as change in basal body temperature (an indication of ovulation) or beginning of the menstrual flow. Even these measures are imperfect, but Parlee points out that many studies don't even bother with independent confirmation of cycle phase. Properly done correlational studies also present problems. Early in their training, scientists learn that correlation gives no information about causation. As Parlee points out, most correlational studies seem to assume that when emotional changes correspond to changes in the menstrual cycle, hormones must be to blame. The model gives priority to biology by implication rather than by direct proof and leaves aside the idea that one's hormone biology might itself change under differing emotional conditions.

Retrospective questionnaires. Even today, one of the most widely used processes for studying premenstrual anxiety and depression is to ask women to fill out a form called the Menstrual Distress Questionnaire (abbreviated MDQ), on which they indicate which among a list of symptoms or moods they recall experiencing at different times of their menstrual cycle. Designed by a physician, Dr. Rudolf Moos, "to find the relationship between a woman's symptoms in different phases of her menstrual cycle," the question-naire lists forty-seven "symptoms," only five of which—all found under the subheading "positive arousal"—have agreeable connota-tions.[34] Table 4.3 sets forth the items on Moos's MDQ. A rapid perusal alerts one to the fact that, starting out with the assumption

TABLE 4.3
*Symptom Scale Groups Made Up from Menstrual,
Premenstrual, Intermenstrual, and Worst Menstrual
Symptom Factor Analyses*

Pain	*Water Retention*
Muscle stiffness	Weight gain
Headache	Skin disorders
Cramps	Painful breasts
Backache	Swelling
Fatigue	
General aches and pains	*Negative Affect*
	Crying
Concentration	Loneliness
Insomnia	Anxiety
Forgetfulness	Restlessness
Confusion	Irritability
Lowered judgment	Mood swings
Difficulty concentrating	Depression
Distractible	Tension
Accidents	
Lowered motor coordination	*Positive Arousal*
	Affectionate
Behavioral Change (i.e., Action)	Orderliness
Lowered school or work performance	Excitement
Take naps, stay in bed	Feelings of well-being
Stay at home	Bursts of energy, activity
Avoid social activities	
Decreased efficiency	*Control Symptoms*
	Feelings of suffocation
Dizziness	Chest pains
Dizziness, faintness	Ringing in the ears
Cold sweats	Heart pounding
Nausea, vomiting	Numbness, tingling
Hot flashes	Blind spots, fuzzy vision

NOTE: Rudolf H. Moos, "Typology of Menstrual Cycle Symptoms," *American Journal of Obstetrics and Gynecology* 103(1969):393.

of menstrual distress, one can design a test to prove one's point. Critics of this work recently designed a Menstrual Joy Questionnaire, which emphasizes positive feelings during the menstrual cycle.[35] The retrospective questionnaire holds additional problems, including the selective memory of the women completing it combined with their prior knowledge of the purpose of the study. Many women grow up with the expectation that they should feel bad just before their periods, and this belief can certainly predispose them to selectively remember feeling bad just before menstruation but not at other times of the month. As one researcher writes in a study on moods and menstruation in college students, "negative behavior exhibited premenstrually is perceived as evidence for the prevailing negative stereotype of female emotional behavior while positive behavior is ignored as something for which biology is irrelevant."[36]

The uncritical use of retrospective studies ignores the twin influences of negative cultural beliefs and outright ignorance about menstruation. A study of forty women at a Michigan clinic serving a poor, multiethnic clientele found patients to be "naive and misinformed about the true function of menstruation."[37] Most (63 percent) had no knowledge of menarche before they began to menstruate, while many thought that "cold" could enter the body during menstruation, "believing that a bath, a shampoo, a walk in the dew or rain might 'back up' menstrual flow and result in a stroke, insanity, or 'quick TB.'"[38]

A second type of investigation shows that middle-class children, presumably well educated about menstruation, still approach menarche with fixed ideas about what to expect. In one recent study most premenarcheal girls and boys of the same age believed menstruation to be a physically and emotionally disruptive event, while another found that 85 percent of the premenarcheal girls studied thought it inappropriate to discuss menstruation with boys.[39] Regardless of class or educational level, large numbers of women grow up with strong negative attitudes toward menstruation. It is obvious to many that these firm beliefs might affect answers not only to retrospective questionnaires but also the physiology of menstruation itself, yet possible effects are ignored in mainstream menstrual-cycle research.

Daily self-reports. Some researchers overcome the problem of selective memory on retrospective forms by using daily self-reports. With this method groups of women keep daily menstrual calendars on which they record their moods and the first day of menstruation. Until very recently, however, such studies have not been "blind"—the subject always knew that she was part of an attempt to correlate mood and menstrual phase.

Thematic analysis. In the fourth type of study reviewed by Parlee, researchers ask women to speak in unstructured fashion about daily life experiences. The interviewers then analyze the language content, scoring for indications of hostility and anxiety. By performing such analyses over the course of one or more menstrual cycles, investigators hope to correlate emotional changes with physiological ones. Although with this approach, the subject may be unaware of the study's purpose, the analyst isn't. In other words, this too is a "sighted" study with all the pitfalls of subjective analysis. The most reliable studies employ double blinds, in which neither the experimental subject nor the person gathering and categorizing the results knows the true purpose of the study. The few studies of premenstrual emotional changes done in this way fail to show cyclical mood alterations.[40]

Parlee's report merely scratches the surface, beneath which lies so huge a number of methodological problems that the thoughtful menstrual-cycle researcher risks paralysis in dealing with them, while they simply overwhelm the unwary layperson. Table 4.4 categorizes major areas of methodological difficulty into subheadings such as proper controls, appropriate use of statistical analyses, hormone measurements, choice of subject sample, the search for valid measures of mood, and overall experimental design, listing within each subheading some of the items researchers need to consider.

How is it that so many different and seemingly obvious things could be wrong with the bulk of menstrual-cycle research? The answer, hidden within the walls of the research structure, is usually plastered over with clever techniques and detailed measurements. Why, for instance, do researchers looking at women's hormone cycles and mood changes fail to mention the monthly cycle of testosterone?[41] The answer, I suggest is that testosterone is seen as a male hormone, hence "normal," and thus not an obvious subject for inquiry when looking at the "abnormalities" of menstrual mood changes. How, too, can so many highly trained researchers fail even to discuss the problem of double-blind experimentation? The answer lies in their belief in biological primacy. One need worry about double blinds only if one remains conscious of the fact that thoughts, mind-sets, and emotions can affect one's physiology. Of course most women know this perfectly well, since overexcitement, exhaustion, travel, illness, and stress can alter the timing of one's period, change the number and intensity of premenstrual signals, and influence the presence or absence of the menstrual flow and its degree of discomfort; these are all variations in the physiological expression of the monthly cycle, influenced by one's emotional state.

Dr. Randi Koeske, author of the most up-to-date analyses of menstrual-cycle theory,[42] paints in clear detail the consequences of reliance on "a normative 'disease-model' framework" for the menstrual cycle, a framework that gives precedence to biological variables

TABLE 4.4

Methodological Considerations in Menstrual-Cycle Research

Controls

How does one establish a baseline of "normality" (needed for a medical model of menstrual emotionality)?

Are men used as controls?

Are women in the study taking birth control pills, which flatten out the monthly hormonal fluctuations?

Have cross-cultural studies been done?

Does the study include large population surveys?

Choice of Subject Sample

How large should the sample be?

How diverse should the sample be in terms of:

age (most studies are done on college women);

race (most studies are done on white women);

religion (some studies show differences between Jewish, Catholic, and Protestant women);

class (the above differences are confounded by socioeconomic class differences);

whether subjects have given birth;

whether subjects volunteer or are randomly chosen;

whether subjects have a history of mental illness (as is the case in a number of frequently cited studies)?

Experimental Design

Is the study double-blind, single-blind, or not blind in either direction?

Is more than one cycle studied to control for monthly variability? (No more than two consecutive cycles are usually studied.)

How is the staging of the cycle assessed: with physiological measures (hormonal or temperature changes)? Subject reporting?

How frequently do subjects report on mood? Four times per cycle? Daily?

Statistical Analyses

Is there an assumption that the menstrual cycle is an independent variable, that is, that it is unchanged by the events under study—mood, stress, and so forth?

Do the authors consider general and specific limitations of the "analysis of variance" approach (in which specific "causes" or covariants are statistically extracted from a multifaceted set of data)?

Is it appropriate to use parametric statistics, designed to deal with populations that fit into a bell-shaped curve, for data with a very high degree of individual variability?

What is the actual magnitude of change observed?

Validation of Mood Measurement

Independent validation of accuracy of psychological mood tests used in menstrual-cycle research has usually been inadequate or nonexistent.

Use of more than one method of assessing mood (as a cross-check) has rarely been done. When done, results are most often inconsistent.

Categories of mood (tension, agitation, depression, anger, and so on) are often merged as if they had the same meaning and effects.

Researchers have looked almost exclusively for *negative* mood changes because of hidden assumptions that no positive changes are associated with the menstrual cycle.

TABLE 4.4 *(continued)*

Researchers have used overly detailed mood measures that may pick up statistical
variations with no actual psychological, emotional, or behavioral impact.

Assessment of noncycle-related mood changes, such as mood rhythms associated
with time of day, day of week, or time of year, may give false results or
interact with (potentiate) menstrually associated mood changes.

Problems with the Meaning of Hormone Measurement

Choice of hormone to measure: virtually no studies mention that testosterone
levels change in a predictable cyclic fashion during the menstrual cycle.

Treatment of hormones as individual, independent variables rather than part of a
complex, interacting physiological system.

Assumption of unidirectional causative chain (hormones dictate behavior), when it
is well known that the reverse is also true: menstrual synchrony, stress and
cycles, nursing, and exercise, for example, all alter the menstrual cycle.

Use of technically sophisticated methods of hormone measurement without
corresponding attention to their biological meaning. For example:

Are there receptors in presumed target organs for hormones found to be
circulating in the blood or excreted in the urine? Only if there *are* can the
hormones exert their physiological effects.

Are hormones in blood plasma bound to protein, which renders them
physiologically inactive, or are they freely circulating? (Many studies do not
make this distinction.)

Is the model inappropriately based on absolute levels of hormone, rather than on
a dynamic model which looks at rates of change over time.

The hormonal basis of cycle control is incompletely understood, especially with
reference to the role of behavior in cycle control.

SOURCES: W. P. Collins and J. R. Newton, "The Ovarian Cycle," in *Biochemistry of Women:
Clinical Concepts,* ed. A. S. Curry and J. V. Hewitt (Boca Raton, Fla.: CRC Press, 1974); A. J.
Dan, E. Graham, and C. P. Beecher, eds., *The Menstrual Cycle,* vol. 1 (New York: Springer,
1980); P. Komnenich et al., eds., *The Menstrual Cycle,* vol. 2 (New York: Springer, 1981); Randi
Koeske, "Premenstrual Emotionality: Is Biology Destiny?," *Women and Health* 1(1976):11–13;
Randi Koeske, "Lifting the Curse of Menstruation: Toward a Feminist Perspective on the
Menstrual Cycle," *Women and Health* 8(1983):1–16; H. Persky, "Reproductive Hormones,
Moods and the Menstrual Cycle," in *Sex Differences in Behavior,* ed. R. C. Friedman, R. M.
Richart, and R. L. Vande Wiele (New York: Krieger, 1974).

(which researchers view as representing the "real" underlying cause
of behavior), while relegating social-cognitive variables to secondary
positions.[43] It is a theoretical approach that emphasizes technical
measurement even when it is devoid of meaning and focuses on
"context-free factors inside the organism," as if the organism
existed separately from its environment; at the same time the
normative disease-model framework views "changeableness, emotionality, and rhythmicity" as either "inherently unhealthy" or as
changes from norms (usually based on men) about how often and
to what degree rhythmic variations are "appropriate." Such a viewpoint allows researchers to ignore the historical and cultural contexts

of their work and, as Koeske so nicely puts it, reduces them to "capturing brief snapshots of time, rather than allowing processes to unfold in context."[44]

Despite such insights, it is not yet time to write the ending to the saga of menstrual-cycle research. One can look forward to new illumination of this subject. A younger generation of scientists has begun to take fresh research initiatives, forging different theoretical approaches and using them to make their way toward better designed, more informative studies of the menstrual cycle. One of the most pleasurable aspects of reading the work of these scientists is that it stems from a full respect for women. A close second is the intellectual challenge encountered in considering complex, contextual research that precludes the reduction of human behavior to some simple biological variable. "The *real* alternative to biomedicine," Koeske writes, "is a system of health research . . . which finds a way to reintegrate the whole person from the jigsaw of parts created by modern scientific medicine. The strength of the feminist perspective is the recognition that the parts biomedicine currently recognizes cannot be reassembled into a whole."[45] The challenge presented to menstrual-cycle researchers is to find the missing pieces and light the way toward integration.

Koeske suggests several steps in that direction which, if followed inventively, will help deepen our understanding of the problem. These include:

1. Start with clearly stated and testable hypotheses, rather than creating a hypothesis after collecting the data.
2. Use more than one method to test the hypothesis, rather than relying solely on one highly modish approach.
3. Listen carefully to what women actually say about their bodies, lives, and perceptions, and treat their views as data, rather than restricting their choice of feelings in order to construct an "objective" measurement.
4. Beware (especially feminists) of replacing a biologically reductive theory with a socially reductive one.[46]

Like the merits of Mom and apple pie, Koeske's cautions may seem self-evident. When applied to actual research problems, however, they bring interesting results. Some examples follow.

1. Dr. Alice Rossi introduced "the social week" into her studies

and found that both men and women experienced cyclical changes during this period. They tended to feel down on Tuesdays and happiest on weekends.[47]

2. Dr. Koeske bases some of her recent work on attribution theory, which tries to understand how people interpret their experiences. In one set of studies she asked students presented with made-up "case studies" that included menstrual-cycle information to interpret the behaviors of hypothetical "patients." Her findings bore out her hypothesis that only hostile or angry behaviors during the premenstruum were attributed to biology; the students believed that positive behaviors during the same period stemmed from individual personality traits or some external positive situation.[48]

3. Dr. Alice Dan, another member of the "new wave" of menstrual-cycle researchers, took up a point first made by Parlee—that we have yet to use two different measures to show the same menstrually related mood changes. In her study Dan compared self-reports of menstrual changes with analysis of open-ended conversations held at different times during the menstrual cycle. Her finding that the two independent assessments disagreed with one another raises some important questions, both about the validity of past studies and the interpretation of present ones. She considers reasonable the "suggestion that we stop measuring 'mood' changes over the menstrual cycle until we have developed a better understanding of what we are measuring."[49]

4. Dr. Sharon Golub found that premenstrually related increases in anxiety were far smaller than the heightened anxiety experienced by students subjected to the stress of an examination. In general she found premenstrually related mood changes to be of small magnitude, concluding that "the premenstrual hormonal changes appear to impose little psychological burden," and are often so slight that women "are sometimes not even aware of them."[50]

One of the lessons of so much critical analysis of menstrual-cycle research is that there is no such thing as a perfect study. The examples just cited may best be thought of as pilot projects. Through them, researchers will learn how to design good studies, how to validate the methods they use, and how to state their hypotheses explicitly and in testable fashion.

Change *does* loom on the horizon. But it will come not from proponents of the medical model of menstruation, from those who suggest that women take hormones, diuretics, or tranquilizers to control their unruly bodies. My optimism stems rather from the number and variety of thoughtful studies, such as those by Rossi, Koeske, Dan, Golub, and others. In the past four years at least three book-length volumes have appeared, each filled with exploratory studies that test the accuracy of old modes of measurement and use new, ingenious, multidimensional experimental approaches, with proper controls and new theoretical frameworks.[51] As Dr. Koeske puts it: "Who knows? With a little luck and a lot of hard work, the only thing that may turn out to be unusual or uninterpretable about female premenstrual emotionality is the fact that no one understood it before."[52] And understanding, I believe, *will* come from the type of research undertaken by the feminist biologists and social scientists of today.

Menopause: The Storm Before the Calm

An unlikely specter haunts the world. It is the ghost of former womanhood (see figure 4.1), "unfortunate women abounding in the streets walking stiffly in twos and threes, seeing little and observing less. . . . The world appears [to them] as through a grey veil, and they live as docile, harmless creatures missing most of life's values." According to Dr. Robert Wilson and Thelma Wilson, though, one should not be fooled by their "vapid cow-like negative state" because "There is ample evidence that the course of history has been changed not only by the presence of estrogen, but by its absence. The untold misery of alcoholism, drug addiction, divorce and broken homes caused by these unstable estrogen-starved women cannot be presented in statistical form.[53]

Rather than releasing women from their monthly emotional slavery to the sex hormones, menopause involves them in new horrors. At the individual level one encounters the specter of sexual degeneration, described so vividly by Dr. David Reuben: "The vagina begins to shrivel, the breasts atrophy, sexual desire disappears. . . . Increased facial hair, deepening voice, obesity . . . coarsened features, enlargement of the clitoris, and gradual baldness complete the tragic picture. Not really a man but no longer a functional

FIGURE 4.1
One Medical View of the Postmenopausal Woman

NOTE: Robert A. Wilson and Thelma A. Wilson, "The Fate of the Nontreated Postmenopausal Woman: A Plea for the Maintenance of Adequate Estrogen from Puberty to the Grave," *Journal of the American Geriatric Society* 11(1963):351, 356. Reprinted with permission from the W. B. Saunders Co. According to Wilson and Wilson, the woman on the left shows "some of the stigmata of 'Nature's defeminization.' The general stiffness of muscles and ligaments, the 'dowager's hump,' and the 'negativistic' expression are part of a picture usually attributed to age alone. Some of these women exhibit signs and symptoms similar to those in the early stages of Parkinson's disease. They exist rather than live." The woman on the right shows the "typical appearance of the desexed woman found on our streets today. They pass unnoticed and, in turn, notice little."

woman, these individuals live in the world of intersex.[54] At the demographic level writers express foreboding about women of the baby-boom generation, whose life span has increased from an average forty-eight years at the turn of the century to a projected eighty years in the year 2000.[55] Modern medicine, it seems, has played a cruel trick on women. One hundred years ago they didn't live long enough to face the hardships of menopause but today their increased longevity means they will live for twenty-five to thirty years beyond the time when they lose all possibility of reproducing. To quote Dr. Wilson again: "The unpalatable truth must be faced that all postmenopausal women are castrates."[56]

But what medicine has wrought, it can also rend asunder. Few publications have had so great an effect on the lives of so many women as have those of Dr. Robert A. Wilson who pronounced menopause to be a disease of estrogen deficiency. At the same time in an influential popular form, in his book *Feminine Forever*, he offered a treatment: estrogen replacement therapy (ERT).[57] During the first seven months following publication in 1966, Wilson's book sold one hundred thousand copies and was excerpted in *Vogue* and *Look* magazines. It influenced thousands of physicians to prescribe estrogen to millions of women, many of whom had no clinical "symptoms" other than cessation of the menses. As one of his credentials Wilson lists himself as head of the Wilson Research Foundation, an outfit funded by Ayerst Labs, Searle, and Upjohn, all pharmaceutical giants interested in the large potential market for estrogen. (After all, no woman who lives long enough can avoid menopause.) As late as 1976 Ayerst also supported the Information Center on the Mature Woman, a public relations firm that promoted estrogen replacement therapy. By 1975 some six million women had started long-term treatment with Premarin (Ayerst Labs' brand name for estrogen), making it the fourth or fifth most popular drug in the United States. Even today, two million of the forty million postmenopausal women in the United States contribute to the $70 million grossed each year from the sale of Premarin-brand estrogen.[58] The "disease of menopause" is not only a social problem: it's big business.[59]

The high sales of Premarin continue despite the publication in 1975 of an article linking estrogen treatment to uterine cancer.[60] Although in the wake of that publication many women stopped taking estrogen and many physicians became more cautious about prescribing it, the idea of hormone replacement therapy remains with us. At least three recent publications in medical journals seriously consider whether the benefits of estrogen might not outweigh the dangers.[61] The continuing flap over treatment for this so-called deficiency diseases of the aging female forces one to ask just what *is* this terrible state called menopause? Are its effects so unbearable that one might prefer to increase, even ever-so-slightly, the risk of cancer rather than suffer the daily discomforts encountered during "the change of life"?

Ours is a culture that fears the elderly. Rather than venerate their years and listen to their wisdom, we segregate them in housing built for "their special needs" separated from the younger generations from which we draw hope for the future. At the same time we allow millions of old people to live on inadequate incomes, in fear that serious illness will leave them destitute. The happy, productive elderly remain invisible in our midst. (One must look to feminist publications such as *Our Bodies, Ourselves* to find women who express pleasure in their postmenopausal state.) Television ads portray only the arthritic, the toothless, the wrinkled, and the constipated. If estrogen really is the hormone of youth and its decline suggests the coming of old age, then its loss is a part of biology that our culture ill equips us to handle.

There is, of course, a history to our cultural attitudes toward the elderly woman and our views about menopause. In the nineteenth century physicians believed that at menopause a woman entered a period of depression and increased susceptibility to disease. The postmenopausal body might be racked with "dyspepsia, diarrhea . . . rheumatic pains, paralysis, apoplexy . . . hemorrhaging . . . tuberculosis . . . and diabetes," while emotionally the aging female risked becoming irritable, depressed, hysterical, melancholic, or even insane. The more a woman violated social laws (such as using birth control or promoting female suffrage), the more likely she would be to suffer a disease-ridden menopause.[62] In the twentieth century, psychologist Helene Deutsch wrote that at menopause "woman has ended her existence as a bearer of future life and has reached her natural end—her partial death—as a servant of the species."[63] Deutsch believed that during the postmenopausal years a woman's main psychological task was to accept the progressive biological withering she experienced. Other well-known psychologists have also accepted the idea that a woman's life purpose is mainly reproductive and that her postreproductive years are ones of inevitable decline. Even in recent times postmenopausal women have been "treated" with tranquilizers, hormones, electroshock, and lithium.[64]

But should women accept what many see as an inevitable emotional and biological decline? Should they believe, as Wilson does, that "from a practical point of view a man remains a man until the end," but that after menopause "we no longer have the 'whole woman'—only the 'part woman'"?[65] What is the real story of menopause?

The Change: Its Definition and Physiology

In 1976 under the auspices of the American Geriatric Society and the medical faculty of the University of Montpellier, the First International Congress on the Menopause convened in the south of

France. In the volume that emerged from that conference, scientists and clinicians from around the world agreed on a standard definition of the words *menopause* and *climacteric.* "Menopause," they wrote, "indicates the final menstrual period and occurs during the climacteric. The climacteric is that phase in the aging process of women marking the transition from the reproductive stage of life to the non-reproductive stage."[66] By consensus, then, the word *menopause* has come to mean a specific event, the last menstruation, while *climacteric* implies a process occurring over a period of years.*

During the menstrual cycle the blood levels of a number of hormones rise and fall on a regular basis. At the end of one monthly cycle, the low levels of estrogen and progesterone trigger the pituitary gland to make follicle stimulating hormone (FSH) and luteinizing hormone (LH). The FSH influences the cells of the ovary to make large amounts of estrogen, and induces the growth and maturation of an oocyte. The LH, at just the right moment, induces ovulation and stimulates certain ovarian cells to form a progesterone-secreting structure called a corpus luteum. When no pregnancy occurs the life of the corpus luteum is limited and, as it degenerates, the lowered level of steroid hormones calls forth a new round of follicle stimulating and luteinizing hormone synthesis, beginning the cycle once again. Although the ovary produces the lion's share of these steroid hormones, the cells of the adrenal gland also contribute and this contribution increases in significance after menopause.

What happens to the intricately balanced hormone cycle during the several years preceding menopause is little understood, although it seems likely that gradual changes occur in the balance between pituitary activity (FSH and LH production) and estrogen synthesis.[68] One thing, however, is clear: menopause does not mean the *absence* of estrogen, but rather a gradual lowering in the availability of *ovarian* estrogen. Table 4.5 summarizes some salient information about changes in steroid hormone levels during the menstrual cycle and after menopause. In looking at the high point of cycle synthesis and then comparing it to women who no longer menstruate the

---

* There is also a male climacteric, which entails a gradual reduction in production of the hormone testosterone over the years as part of the male aging process. What part it plays in that process is poorly understood and seems frequently to be ignored by researchers, who prefer to contrast continuing male reproductive potency with the loss of childbearing ability in women.[67]

TABLE 4.5

Hormone Levels as a Percentage of Mid-Menstrual-Cycle High Point

	Type of Estrogen			Proges-terone	Testos-terone	Andro-stene-dione
	Estrone	Estradiol	Estriol			
Premenopausal						
Stage of Menstrual Cycle						
Early (menses)	20%	13%	67%	100%	55%	87%
Mid (ovulation)	100	100	—	—	100	100
Late (premenstrual)	49	50	100	—	82	—
Postmenopausal	17	3	50	50	23	39

SOURCE: Wulf H. Utian, *Menopause in Modern Perspectives* (New York: Appleton-Century-Crofts, 1980), 32.

most dramatic change is seen in the estrogenic hormone estradiol.* The other estrogenic hormones, as well as progesterone and testosterone, drop off to some extent but continue to be synthesized at a level comparable to that observed during the early phases of the menstrual cycle. Instead of concentrating on the notion of estrogen deficiency, however, it is more important to point out that: (1) postmenopausally the body makes different kinds of estrogen; (2) the ovaries synthesize less and the adrenals more of these hormones; and (3) the monthly ups and downs of these hormones even out following menopause.

While estrogen levels begin to decline, the levels of FSH and LH start to increase. Changes in these hormones appear as early as eight years before menopause.[69] At the time of menopause and for several years afterward, these two hormones are found in very high concentrations compared to menstrual levels (FSH as many as fourteen times more concentrated than premenopausally, and LH more than three times more). Over a period of years such high levels are reduced to about half their peak value, leaving the postmenopausal woman with one-and-one-half times more LH and seven times more FSH circulating in her blood than when she menstruated regularly.

It is to all of these changes in hormone levels that the words such as *climacteric* and *menopause* refer. From these alterations Wilson and others have chosen to blame estrogen for the emotional

---

* Estrogens are really a family of structurally similar molecules. Their possibly different biological roles are not clearly delineated.

deterioration they believe appears in postmenopausal women. Why they have focused on only one hormone from a complex system of hormonal changes is anybody's guess. I suspect, however, that the reasons are (at least) twofold. First, the normative biomedical disease model of female physiology—the same one criticized by Koeske when applied to menstrual research—looks for simple cause and effect. Most researchers, then, have simply assumed estrogen to be a "cause" and set out to measure its "effect." The model or framework out of which such investigators work precludes an interrelated analysis of all the different (and closely connected) hormonal changes going on during the climacteric. But why single out estrogen? Possibly because this hormone plays an important role in the menstrual cycle as well as in the development of "feminine" characteristics such as breasts and overall body contours. It is seen as the quintessential female hormone. So where could one better direct one's attention if, to begin with, one views menopause as the loss of true womanhood?

Physical changes do occur following menopause. Which, if any, of these are caused by changing hormone levels is another question. Menopause research comes equipped with its own unique experimental traps.[70] The most obvious is that a postmenopausal population is also an aging population. Do physical and emotional differences found in groups of postmenopausal women have to do with hormonal changes or with other aspects of aging? It is a difficult matter to sort out. Furthermore, many of the studies on menopause have been done on preselected populations, using women who volunteer because they experience classic menopausal "symptoms" such as the hot flash. Such investigations tell us nothing about average changes within the population as a whole. In the language of the social scientist, we have no baseline data, nothing to which we can compare menopausal women, no way to tell whether the complaint of a particular woman is typical, a cause for medical concern, or simply idiosyncratic.

Since the late 1970s feminist researchers have begun to provide us with much-needed information. Although their results confirm some beliefs long held by physicians, these newer investigators present them in a more sophisticated context. Dr. Madeleine Goodman and her colleagues designed a study in which they drew information from a large population of women ranging in age from thirty-five to sixty. All had undergone routine multiphasic screening at a health maintenance clinic, but none had come for problems

concerning menopause. From the complete clinic records they selected a population of women who had not menstruated for at least
one year and compared their health records with those who still
menstruated, looking at thirty-five different variables, such as cramps,
blood glucose levels, blood calcium, and hot flashes, to see if any of
these symptoms correlated with those seen in postmenopausal
women. The results are startling. They found that only 28 percent
of Caucasian women and 24 percent of Japanese women identified
as postmenopausal "reported traditional menopausal symptoms such
as hot flashes, sweats, etc., while in non-menopausal controls, 16%
in Caucasians and 10% in Japanese also reported these same symptoms."[71] In other words, 75 percent of menopausal women in their
sample reported no remarkable menopausal symptoms, a result in
sharp contrast to earlier studies using women who identified themselves as menopausal.

In a similar exploration, researcher Karen Frey found evidence
to support Goodman's results. She wrote that menopausal women
"did not report significantly greater frequency of physical symptoms
or concern about these symptoms than did pre- or post-menopausal
women."[72] The studies of Goodman, Frey, and others[73] draw into
serious question the notion that menopause is generally or necessarily
associated with a set of disease symptoms. Yet at least three physical
changes—hot flashes, vaginal dryness and irritation, and osteoporosis—and one emotional one—depression—remain associated in
the minds of many with the decreased estrogen levels of the
climacteric. Goodman's work indicates that such changes may be
far less widespread than previously believed, but if they are trou-
blesome to 26 percent of all menopausal women they remain an
appropriate subject for analysis.

We know only the immediate cause of hot flashes: a sudden
expansion of the blood flow into the skin. The technical term to
describe them, *vasomotor instability*, means only that nerve cells
signal the widening of blood vessels allowing more blood into the
body's periphery. A consensus has emerged on two things: (1) the
high concentration of FSH and LH in the blood probably causes hot
flashes, although exactly how this happens remains unknown; and
(2) estrogen treatment is the only currently available way to suppress
the hot flashes. One hypothesis is that by means of a feedback
mechanism, artificially raised blood levels of estrogen signal the
brain to tell the pituitary to call off the FSH and LH. Although
estrogen does stop the hot flashes, its effects are only temporary;

remove the estrogen and the flashes return. Left alone, the body eventually adjusts to the changing levels of FSH and LH. Thus a premenopausal woman has two choices in dealing with hot flashes: she can either take estrogen as a permanent medication, a course Wilson refers to as embarking "on the great adventure of preserving or regaining your full femininity,"[74] or suffer some discomfort while nature takes its course. Since the longer one takes estrogen the greater the danger of estrogen-linked cancer, many health-care workers recommend the latter.[75]

Some women experience postmenopausal vaginal dryness and irritation that can make sexual intercourse painful. Since the cells of the vaginal wall contain estrogen receptors it is not surprising that estrogen applied locally or taken in pill form helps with this difficulty. Even locally applied, however, the estrogen enters into the bloodstream, presenting the same dangers as when taken in pill form. There are alternative treatments, though, for vaginal dryness. The Boston Women's Health Collective, for example, recommends the use of nonestrogen vaginal creams or jellies, which seem to be effective and are certainly safer. Continued sexual activity also helps—yet another example of the interaction between behavior and physiology.

Hot flashes and vaginal dryness are the *only* climacteric-associated changes for which estrogen unambiguously offers relief. Since significant numbers of women do not experience these changes and since for many of those that do the effects are relatively mild, the wisdom of ERT must be examined carefully and on an individual basis. Both men and women undergo certain changes as they age, but Wilson's catastrophic vision of postmenopausal women—those ghosts gliding by "unnoticed and, in turn, notic[ing] little"[76]—is such a far cry from reality that it is a source of amazement that serious medical writers continue to quote his work.

In contrast to hot flashes and vaginal dryness, osteoporosis, a brittleness of the bone which can in severe cases cripple, has a complex origin. Since this potentially life-threatening condition appears more frequently in older women than in older men, the hypothesis of a relationship with estrogen levels seemed plausible to many. But as one medical worker has said, a unified theory of the disease "is still non-existent, although sedentary life styles, genetic predisposition, hormonal imbalance, vitamin deficiencies, high-protein diets, and cigarette smoking all have been implicated."[77]

Estrogen treatment seems to arrest the disease for a while, but may lose effectiveness after a few years.[78]

Even more so than in connection with any physical changes, women have hit up against a medical double bind whenever they have complained of emotional problems during the years of climacteric. On the one hand physicians dismissed these complaints as the imagined ills of a hormone-deficient brain, while on the other they generalized the problem, arguing that middle-aged women are emotionally unreliable, unfit for positions of leadership and responsibility. Women had two choices: to complain and experience ridicule and/or improper medical treatment, or to suffer in silence. Hormonal changes during menopause were presumed to be the cause of psychiatric symptoms ranging from fatigue, dizziness, irritability, apprehension, and insomnia to severe headaches and psychotic depression. In recent years, however, these earlier accounts have been supplanted by a rather different consensus now emerging among responsible medical researchers.

To begin with, there are no data to support the idea that menopause has any relationship to serious depression in women. Postmenopausal women who experience psychosis have almost always had similar episodes premenopausally.[79] The notion of the hormonally depressed woman is a shibboleth that must be laid permanently to rest. Some studies have related irritability and insomnia to loss of sleep from nighttime hot flashes. Thus, for women who experience hot flashes, these emotional difficulties might, indirectly, relate to menopause. But the social, life history, and family contexts in which middle-aged women find themselves are more important links to emotional changes occurring during the years of the climacteric. And these, of course, have nothing whatsoever to do with hormones. Quite a number of studies suggest that the majority of women do not consider menopause a time of crisis. Nor do most women suffer from the so-called "empty nest syndrome" supposedly experienced when children leave home. On the contrary, investigation suggests that women without small children are less depressed and have higher incomes and an increased sense of well-being.[80] Such positive reactions depend upon work histories, individual upbringing, cultural background, and general state of health, among other things.

In a survey conducted for *Our Bodies, Ourselves*, one which in no sense represents a balanced cross section of U.S. women, the

Boston Women's Health Collective recorded the reactions of more than two hundred menopausal or postmenopausal women, most of whom were suburban, married, and employed, to a series of questions about menopause. About two-thirds of them felt either positively or neutrally about a variety of changes they had undergone, while a whopping 90 percent felt okay or happy about the loss of childbearing ability![81] This result probably comes as no surprise to most women, but it flies in the face of the long-standing belief that women's lives and emotions are driven in greater part by their reproductive systems.

No good account of adult female development in the middle years exists. Levinson,[82] who studied adult men, presents a linear model of male development designed primarily around work experiences. In his analysis, the male climacteric plays only a secondary role. Feminist scholars Rosalind Barnett and Grace Baruch have described the difficulty of fitting women into Levinson's scheme: "It is hard to know how to think of women within this theory—a woman may not enter the world of work until her late thirties, she seldom has a mentor, and even women with life-long career commitments rarely are in a position to reassess their commitment pattern by age 40," as do the men in Levinson's study.[83]

Baruch and Barnett call for the development of a theory of women in their middle years, pointing out that an adequate one can emerge only when researchers set aside preconceived ideas about the central role of biology in adult female development and listen to what women themselves say. Paradoxically, in some sense we will remain unable to understand more about the role of biology in women's middle years until we have a more realistic *social* analysis of women's postadolescent psychological development. Such an analysis must, of course, take into account ethnic, racial, regional, and class differences among women, since once biology is jettisoned as a universal cause of female behavior it no longer makes sense to lump all women into a single category.

Much remains to be understood about menopause. Which biological changes, for instance, result from ovarian degeneration and which from other aspects of aging? How does the aging process compare in men and women? What causes hot flashes and can we find safe ways to alleviate the discomfort they cause? Do other aspects of a woman's life affect the number and severity of menopausally related physical symptoms? What can we learn from studying the experience of menopause in other, especially non-Western cultures? A number of researchers have proposed effective ways of finding answers to these questions.[84] We need only time, research dollars, and an open mind to move forward.

Conclusion

The premise that women are by nature abnormal and inherently diseased dominates past research on menstruation and menopause. While appointing the male reproductive system as normal, this viewpoint calls abnormal any aspect of the female reproductive life cycle that deviates from the male's. At the same time such an analytical framework places the essence of a woman's existence in her reproductive system. Caught in her hormonal windstorm, she strives to attain normality but can only do so by rejecting her biological uniqueness, for that too is essentially deformed: a double bind indeed. Within such an intellectual structure no medical research of any worth to women's health can be done, for it is the blueprint itself that leads investigators to ask the wrong questions, look in the wrong places for answers, and then distort the interpretation of their results.

Reading through the morass of poorly done studies on menstruation and menopause, many of which express deep hatred and fear of women, can be a discouraging experience. One begins to wonder how it can be that within so vast a quantity of material so little quality exists. But at this very moment the field of menstrual-cycle research (including menopause) offers a powerful antidote to that disheartenment in the form of feminist researchers (both male and female) with excellent training and skills, working within a new analytical framework. Rejecting a strict medical model of female development, they understand that men and women have different reproductive cycles, *both* of which are normal. Not binary opposites, male and female physiologies have differences *and* similarities. These research pioneers know too that the human body functions in a social milieu and that it changes in response to that context. Biology is not a one-way determinant but a dynamic component of our existence. And, equally important, these new investigators have learned not only to *listen* to what women say

121

about themselves but to *hear* as well. By and large, these researchers are not in the mainstream of medical and psychological research, but we can look forward to a time when the impact of their work will affect the field of menstrual-cycle research for the better and for many years to come.

If women are seen as emotional slaves of their reproductive physiologies, men do not get off scot-free: *their* "problem hormone" is testosterone. In normal amounts it is seen as the source of many positive traits—drive, ambition, success. But too high a quantity, so the theory goes, can cause antisocial behavior—crime, violence, and even war. In contrast, women's lower testosterone levels mean they are less likely to make it in the world of the hard-driving professional, although they can derive pride from their more peaceable manner. This tale of testosterone and aggression forms yet another subplot in the collection of biological stories about male and female behavior. The chapter that follows analyzes its text.

5

HORMONES AND AGGRESSION: AN EXPLANATION OF POWER?

We are assuming . . . that there are no differences
between men and women except on the hormonal
system that renders the man more aggressive. This
alone would explain patriarchy, male dominance,
and male attainment of high-status roles; for the
male hormonal system gives men an insuperable
"head start."   —STEVEN GOLDBERG, 1973

Of all the vulgar modes of escaping from the
consideration of the effect of social and moral
influences upon the human mind, the most vulgar is
that of attributing the diversities of conduct and
character to inherent natural differences.
                    —JOHN STUART MILL (1806–1873)

THE YEAR: 1969. The place: a crowded meeting room. The
event: a speech (which I have just completed) about sex discrimination and equality for women. The questions and comments begin,
and I field them deftly until a young man stands up.

"What about rats?" he asks with a self-satisfied smile. "*Everyone*
knows that male rats are more aggressive. That's because they have
more testosterone. It's scientifically proven. Women aren't treated
unequally. They can't compete because they just don't have enough
male hormones."

I am stunned. My thought processes screech to a halt. While I
mutter something about the inapplicability of animal studies to

123

human behavior, my brain searches itself for information. "What *about* rats," I puzzle. "And who gives a damn?"

As it turns out, lots of people give a damn. Take, for example, George Gilder, an unofficial spokesperson for Ronald Reagan's economic policies. In an article entitled "The Case Against Women in Combat,"[1] he writes, "The hard evidence is overwhelming that men are more aggressive, competitive, risk-taking . . . [and] more combative than women." And he worries that bringing equal rights straight to the core of the army may put the United States in mortal danger, pointing out that the USSR not only has a larger army, it also has one that is all male. "It would be unfortunate," he frets, "for American leaders to give the impression that they regard combat chiefly as an obstacle to women's rights." Earlier in this century feminists gave the same theme—that of male bellicosity— a reverse twist. If women got the vote, some suffragists believed, there would be an end to war. When women entered politics their natural peaceableness would prevail, and men, unable without women's help to control their primal destructive urges, would finally join the ranks of the civilized.

The theme of male strength and female weakness is not new. In the nineteenth century authorities argued that man, naturally stronger, "is fitted for . . . civil and political employments" while "the consciousness of her physical weakness renders women timid and sedentary. . . . Woman . . . fit only for sedentary occupations . . . necessarily remains much in the interior of the house."[2] One hundred and fifty years later Steven Goldberg, in his comprehensive statement on "the inevitability of patriarchy," continues to use the theory of male aggression as a key to understanding social structure. Males, he claims, have a little biological edge, a little extra aggression, which makes them want to excel. If women seek to alter this fact of life, to change the political system of male domination, he warns that they inevitably will fail.

> Women follow their own physiological imperatives. . . . In this and every other society [men] look to women for gentleness, kindness, and love, for refuge from a world of pain and force. . . . In every society a basic male motivation is the feeling that the women and children must be protected . . . the feminist cannot have it both ways: if she wishes to sacrifice all this, all that she will get in return is the right to meet men on male terms. *She will lose.*[3] [Emphasis added]

In other words, if the rats are telling us the truth, sex discrimination doesn't exist; domination and hierarchy result naturally from our animal nature, and women should get out of the business of competing in a man's world, a world in which their inferiority will out. One psychologist, Dr. Corinne Hutt, explicitly connects aggression to ambition and drive, on the one hand pointing to "androgens as the *fons et origo* of all drive" and on the other, suggesting that women's lack of competitive striving "militates against great achievements in competitive fields" such as science.[4]

According to sociobiologists, humans are biologically determined to be aggressive in times of scarcity. The evidence? To quote Dr. David Barash: "In the American South, whenever the price of cotton fell, the frequency of lynchings would rise." Aggression, he argues, is a mode of competition "often utilized when it proves adaptive. . . . Demogogues . . . have used [this] to manipulate human behavior. When times are hard people are eager to take out their aggression on . . . scapegoats. . . . This was the fate of the Jews in Nazi Germany and blacks in the American South."[5]

Barash argues further that there is a biological basis to warfare, which he views as nothing more than organized group aggression; he agrees, too, with Goldberg's contention that greater male aggressiveness leads men to occupy politically dominant roles ("the exclusion of women from major policy roles is an international, species-wide phenomenon") and, he writes: "There seems to be a long evolutionary history behind this pattern of rivalry and tension between maturing males and the established male hierarchy." This natural struggle of young males to break into the power circles occupied by older ones has its inevitable consequences: "It makes good biological sense that juvenile delinquency is more common among boys than among girls and that men are more prone than women to commit violent crimes. . . . Automobile insurance is also revealing, with the highest rates being charged to unmarried male drivers under twenty-five. . . . The rates . . . reflect, too, the predictions of evolutionary biology."[6]

Barash's implicit argument goes something like this: male hormones evolved as physiological messengers that control aggression. These hormones, in response to particular sorts of stimuli, evoke a wide range of behavior: social dominance, criminal activity, reckless driving, lynchings, genocide, and even modern warfare. It is difficult to get a firm argumentative grip on these claims, because they take a wide variety of phenomena, which have only the most tenuous and superficial interconnections, and lump them together under the

125

flag of natural biological causation. In the process of blending such diverse human activities, all analytical power is lost. The idea of opposing political forces in the world, the idea that the Great Depression in Europe might have been connected in some manner to the rise of nazism, the fact that lynchings in the South decreased as the result of an intensely political anti-lynching campaign, all these and of course much more, become epiphenomena—backwater by-products of our underlying hormonal drives. Far from offering the tradition of scientific analysis (taking apart in order to understand), the type of arguments made by Barash, Goldberg, Hutt, and others constitute a sort of quasi-scientific gruel, a mushing together of different levels of human activity until they become indistinguishable from one another—impossible to sort out and impossible to act upon.

And yet the arguments are very seductive. The idea that male hormones make men more competitive, better at sports, go-getters in the business world, and ready to fight to defend their honor and family certainly captures the popular imagination. Could it be that there is scientific evidence to substantiate such claims?

Male Hormones and the Human Condition

Aggression, violence, crime, riots, war: we may owe all these and more to that simple cholesterol-like molecule called testosterone. Or so some would have us think.[7] Even in this century, many societies, including our own, have tested this belief by castrating men who have a history of violent or antisocial behavior. Only a few investigators, however, have bothered to look at the effectiveness of this draconian approach to crime. Several studies agree that castration reduces the sexual activity of habitual child molesters and rapists. Since the testes produce testosterone, which in turn contributes to sexual potency, this is hardly a surprising outcome. One of the only follow-up studies that looked at the effect of castration on aggression and violence, however, found the treatment so ineffective that nine out of sixteen men castrated solely to "calm them down" subsequently died as a result of aggressive encounters. Doctors no longer use surgical castration to control sex offenders, preferring instead a synthetic female hormone (medroxyprogesterone) to lower testosterone levels in male sex offenders. The few studies on the effects of such "chemical castration" on violent or aggressive behavior[8] conclude that the procedure is not particularly effective.[9] Investigations done on castrated rhesus monkeys also fail to find any straightforward relationship between castration and the lessening of aggressive behavior.[10]

There are remarkably few studies (perhaps no more than six) that attempt to measure aggressive or hostile behavior in humans while at the same time correlating such behavior with testosterone levels in the blood; they are summarized in table 5.1. In principle such experiments divide populations of men into aggressive (hostile) and nonaggressive individuals, measure the levels of testosterone of individuals in each category, and, using statistical tests, look to see whether highly aggressive men have higher testosterone levels circulating in their blood than do less aggressive men. A positive correlation does not necessarily mean that testosterone causes aggression. Elevated testosterone levels may, in fact, *result from* aggressive behavior. Two studies on populations of prisoners reach opposite conclusions, one claiming a positive correlation between testosterone levels and record of aggressive acts, and the other finding no correlation. Three additional studies try to correlate scores on psychological tests with testosterone levels. Although one of these finds a relationship between high hostility scores and testosterone, the other two were unable to repeat the finding. Thus the weight of the evidence, such as it is, suggests no reliable correlation. One worker in the field summarized the available work in the following manner: "If there are significant correlations between aggression and testosterone in humans, we need much better methods of assessing various aspects of behavior than are currently available."[11]

Careful examination of these studies points to a fundamental difficulty that plagues all such research—how can aggression be measured and defined? In the case of prison studies a record of violent behavior seems a plausible way to identify aggressive men. Yet in at least one study, researchers included attempts to escape from prison or from juvenile detention schools in their compilation of violent events in an individual's record.[12] In so doing they throw the validity of their study into question, since it is unclear whether escape attempts are motivated by antisocial aggression or simply by a healthy sense of self-preservation.

TABLE 5.1
Studies of Aggressive Behavior or Feelings and Testosterone Levels in Human Males

Population	Method	Findings
18 normal men, aged 17–28 15 normal men, aged 31–66	Subjects studied for 2 hours, during which testosterone production was measured and several psychological tests designed to measure hostile feelings, depression, and aggressive feelings given.	Correlation between (high) aggression score on test and testosterone production in the younger group but not in the older.[a]
10 prisoners with records of aggressive behavior 11 prisoners jailed for nonviolent crimes	Testosterone levels measured during 2-week period and battery of psychological tests given.	No correlation between fighting behavior or test scores and testosterone levels. Correlation with high testosterone levels and arrests during adolescence.[b]
24 college students	Half scored high on psychological tests of hostility, and half scored low; testosterone production measured while taking tests.	No difference found between the two groups.[c]
36 prisoners aged 18–45	Divided into 3 groups of 12, based on experimenters' knowledge of the individuals and their prison records. Group 1 was very aggressive. Group 2 was not aggressive but was dominant in prison hierarchy. Group 3 was neither aggressive nor dominant. Gave psychological tests and measured testosterone levels.	Aggressive males had statistically higher testosterone levels but no correlation between high scores on tests and testosterone levels.[d]
12 psychiatric inpatients	Weekly testosterone measures for 8 weeks.	No correlation with either aggressive or agitated behavior.[e]
20 healthy volunteers	Took psychological tests and measured testosterone every other morning for 2 months.	No consistency in individuals between hormone levels and high scores for aggression or hostility.[f]

[a] H. Persky, K. D. Smith, and G. K. Basu, "Relation of Psychologic Measures of Aggression and Hostility to Testosterone Production in Man," *Psychosomatic Medicine* 33(1971):265–77.

[b] L. E. Kreuz and R. M. Rose, "Assessment of Aggressive Behavior and Plasma Testosterone in a Young Criminal Population," *Psychosomatic Medicine* 34(1972):321.

[c] H. F. L. Meyer-Bahlburg, et al., "Aggressiveness and Testosterone Measures in Man," *Psychosomatic Medicine* 36(1974):267–74.

[d] Arthur Kling, "Testosterone and Aggressive Behavior in Men and Non-human Primates," in *Hormonal Correlates of Behavior*, ed. B. Eleftheriori and R. Spot (New York: Plenum, 1975), 305–23.

[e] R. M. Rose, "Testosterone, Aggression and Homosexuality: A Review of the Literature and Implications for Future Research," in *Topics in Psychoendocrinology*, ed. E. J. Sachar (New York: Grune and Stratton, 1975).

[f] C. H. Doering, et al., "Plasma Testosterone Levels and Psychologic Measures in Men Over a 2-Month Period," in *Sex Differences in Behavior*, ed. R. C. Friedman, R. M. Richart, and R. L. Vande Wiele (New York: Wiley, 1974).

HORMONES AND AGGRESSION

Many psychologists believe that depression masks angry or aggressive feelings. Some of the studies cited use this idea to measure aggression indirectly by administering psychological tests designed to assess depression. Others use word-association tests that can pick out individuals who have hostile or aggressive feelings. With these tests experimenters attempt to correlate hormone levels and feelings, even though the feelings may never translate into action. Thus in evaluating studies on hormones and human aggression, one must look carefully to see how aggression is defined—is it action or thought, hostility, anger or depression?

Endemic to writings on the biology of aggression is the habit of confusing or interchanging social and biological concepts. Hutt's statement relating individual aggression to ambition provides an example involving psychological concepts.[13] Many psychologists define aggression as the intent or attempt to inflict harm on another person, but *ambition* has a very different meaning. In fact, we usually consider it a positive social quality having to do with one's personal goals for the future. Hutt, though, argues that since androgens (a biological phenomenon) affect aggression (a psychological characteristic) and aggression is related to ambition (a socially accepted attribute), then male hormones must account for the supposedly more ambitious male. The slippage in meaning gets worse when we look at writing about war. One review article in a respected biological journal, for instance, synonymously uses the terms *aggression* and *war*.[14] War is a form of social aggression but has little if anything to do with individual aggression. Traditionally in the United States we have had to draft young men to fill our wartime armies, and many have fled or gone to jail rather than go to war. Warfare is one part of a spectrum of political processes, none of which can be fruitfully analyzed by hormone measurements.

By now I am secretly hoping that some impatient reader will have begun to wonder why I have not yet mentioned either studies on women or studies comparing men and women. That's because there are none. Well—not to exaggerate—there is *one*, published in 1980, which examines testosterone levels in women in different occupations. An anthropologist, Frances Purifoy, and a statistician, Lambert Koopmans, measured the levels of testosterone and androstenenedione (a biochemical precursor of testosterone) in professional and managerial women. They compared these levels to those found in clerical workers, service workers, housewives, and students. Their basic finding, which was often but not always statistically significant, was that students and professional or managerial women have somewhat higher concentrations of testosterone and andros-tenenedione than do clerical and service workers and housewives.[15]

A writer such as Steven Goldberg could argue that the finding that professional and managerial women have higher "male" hormone levels than do housewives proves that male hormones are a positive aid in achieving success. Therefore, he might claim, women with naturally higher testosterone levels are more assertive, more masculine, and more likely to achieve in a male arena. That, however, is *not* the conclusion reached by Purifoy and Koopmans. Instead they point to a well-established observation: stress lowers testosterone levels. This is true for physical stress such as surgery,[16] as well as for psychological stress such as men encounter in basic army training or under threat of attack.[17] Purifoy and Koopmans make the following points:

> Working married women complain of fewer symptoms [of] . . . stress (i.e., headaches, insomnia, depression, anxiety) than housewives. . . . Among those women who work outside the home, traditional women's jobs (i.e., clerical, sales and service work) are the most stressful for women in terms of mental health and heart disease.[18]

They suggest that women in traditionally female occupations experience greater stress, and that this stress causes a reduction in androgen production!

We have thus illustrated the first of two facts that make it impossible to prove straightforward cause-and-effect relationships even when there is a positive correlation between achievement or aggressive behavior and hormone levels. Simply put, correlation does not necessarily imply causation. In the study on women in various occupations, for example, there is good reason to argue that the social situation (the type of work) changed the biological condition (hormone level), rather than the other way around.

There is a second problem with studies that correlate behavior with a single hormone state, and it also has to do with the complexity of human physiology. Mired in the morass of arguments about testosterone and aggressive behavior, it is easy to forget that our bodies have a number of different hormone systems, all of which interact with one another. These are often called neuroendocrine systems because the brain, in response to some external event, activates them by sending messages from the brain cells through the bloodstream to hormone-producing glands. For example, people sometimes say, in referring to a situation that made them angry, "that really got the adrenalin flowing." What they meant, of course, was that their brains translated their psychological state of anger into a chemical message which traveled through the blood-stream to their adrenal glands. The glands responded by producing adrenalin, which in turn enabled people, quite literally (because of the dilated blood vessels), to get "hot under the collar," to become agitated and excited, to scream loudly—that is, to act angrily. The same hormone, adrenalin, can also help give our bodies that extra push necessary to run a little faster and farther when being chased. Under stress, during a fight, when angry, when engaging in behavior some might label aggressive, many different hormone levels change in the body. Thus, to attribute a change in behavior to a change in a single hormone, when many different hormones rise and fall simultaneously, misrepresents the actual physiological events.

A physician named John Mason used rhesus monkeys to study hormonal responses to stress. He trained the monkeys to avoid an electric shock to the foot by pressing a bar with their hands whenever they saw a red light go on in their cages. He then measured hormone levels under three different conditions: (1) before flashing light; (2) with the red light flashing, on a series of days; and (3) again, during a subsequent period, when no light shone. Mason found that, under the stress induced by avoiding the electric shock, cortisone, adrenalin, noradrenalin, thyroid hormone, and growth hormone all increased, while the hormones insulin, testosterone, estradiol, and estrone all decreased. In other words, stress simulta-neously affected many different hormone systems, causing some to increase in activity and others to decrease. From his observations Mason concluded: "The effects . . . are dependent . . . not upon the absolute level of any single hormone, but rather upon the relative 'overall' balance between all participating hormones."[19] In other words, from a physiological point of view, it makes little sense to measure a single hormone level—out of its hormonal context—and use that measurement to draw conclusions about a causal relationship with a particular behavior.*

During the 1960s in Chicago, a man named Richard Speck murdered seven Philippino nursing students brutally and in cold

---

* I am not referring here to well-defined animal studies of reproductive behavior where what is measured is an actual part of the sex act. Obviously, the physiological capacity to reproduce is controlled in part by sex hormones, but this is very limited behavior of a sort quite different from social behaviors such as aggression.

blood. After his arrest, a member of the press reported (wrongly, as it turned out) that Speck had an extra Y chromosome, that his genotype was XYY. A flurry of popular articles on the "XYY syndrome" followed, all of them suggesting that if one Y (in the normal male) causes aggressive behavior, two Y's a criminal doth make. Although the press's infatuation with the extra Y chromosome eventually blew over, to this day, more than twenty years later, I encounter students who have heard about the so-called criminal chromosome. They have only rarely heard, however, that members of the scientific community no longer believe that an extra Y chromosome makes men more violent. Generally, extra chromosomes prevent normal development, which usually results in lowered intelligence. Men with an extra Y (as well as ones with an extra X) score more poorly on IQ tests than do XY males. Most scientists now believe that it is lowered intelligence and *not* genetically induced aggression that correlates with criminal behavior.[20]

The tenuous nature of the data linking an extra Y chromosome to aggressive behavior did not prevent the publication of a large number of research reports (at least thirty-four different articles published during the late sixties and early seventies) attempting to link changes in hormone levels, especially "male" hormones, with the XYY syndrome. Failing to find any consistent relationship, however, one review of the literature concluded, in understated fashion, that "it is quite possible that the . . . behavior disorder encountered in many XYY males is totally unrelated to androgens."[21]

What, then, *do* we know about testosterone levels and adult aggression? Not much. As we have seen, the few existing studies are inconclusive, either because of poor design or questionable definitions of aggression, or because inadequate thought has been given to the difficulty of deciding whether a particular testosterone concentration causes aggression or vice versa. In my view, it is not even sensible to ask whether testosterone causes aggression. But still scientists have not done with it. The most recent work has indeed retreated from the suggestion that adult testosterone levels relate to aggression. Instead a more subtle concept has appeared— the idea of prenatal hormonal conditioning. The press has hopped right onto the bandwagon. A 1981 article ascribes the following comment to one scientist: "The different kinds of behavior that young male monkeys display are completely, scientifically, and uniquely determined by the endocrine conditions that exist before birth."[22]

Aggression Begins at Home (in the Womb)

What happens in the womb? Recall that early in development male embryos make testosterone and dihydrotestosterone, both of which direct the development of the internal and external genitalia. In some animals testosterone also affects brain development. For example, the region of the brain that governs the courtship song of certain male songbirds grows to full size under the influence of male hormones. It is a well-established fact that hormones directly influence brain development for a small number of vertebrates, all species that have a relatively fixed relationship between hormones and certain types of behavior. Many scientific articles rely on these studies in lower vertebrates to bolster the claim that in humans testosterone present during embryonic development predisposes boys to show greater physical activity, to engage in more rough-and-tumble play, and to learn aggressive behavior as they grow to adulthood.

The evidence in humans comes from the study of children exposed to abnormal levels of hormones in utero, about whom scientists asked three related questions. Are boys who are exposed in utero to abnormally high androgen levels more aggressive than normal? Are girls exposed in utero to androgen more aggressive? Are boys exposed in utero to estrogen less aggressive? In sum, does prenatal testosterone augment aggression or estrogen lessen it? As is always the case with human studies, one does not have the luxury of designing a perfectly controlled experiment, but must rely instead on the examination of medical accidents. Mishaps relevant to the study of aggression include children born from mothers who took hormone injections or pills during pregnancy and children suffering from adrenogenital syndrome (AGS), which causes the adrenal glands to produce too much androgen. We have already discussed studies (see chapter 3) suggesting both that AGS children develop a gender identity consistent with the sex of rearing and that the increased androgen levels do not improve intelligence (see chapter 2).

One article, written in 1974, reviews the evidence on fetal

133

hormones, the brain, and sex differences and concludes:

> The data presented clearly compel the conclusion that a fetal hormonal effect is influential in the subsequent development of sex differences.[23]

But another, written in 1979, states:

> It is difficult to attach much clinical significance to many of the studies purporting to demonstrate an effect of . . . prenatal sex hormone . . . on sexually dimorphic behavior. . . . They represent small numbers of patients, often treated with multiple medications for whom suitable controls are difficult to obtain.[24]

Is there truth to be found amidst the fray?

John Money and Anke Ehrhardt first popularized their studies of fetally androgenized girls in their 1972 book, *Man and Woman, Boy and Girl*.[25] They reported on fifteen adrenogenital syndrome (AGS) girls and ten girls whose mothers received progestins during pregnancy, a treatment that masculinized the external genitalia of otherwise normal XX girls. Although the progestin-exposed girls required no additional medication, the children with AGS needed continual cortisone treatment in order for their adrenal glands to function normally. All twenty-five girls had operations to change the appearance of their masculinized genitalia. Money and Erhardt never made clear, except through one photograph, that such "correction" to what they term a "normal female appearance" involved clitoridectomy.[26] For comparison, they talked with the same number of medically normal girls of the same age and socioeconomic background. Although they interviewed all fifty girls and their mothers, they made no first-hand observations of behavior. In contrast with the control group they found:

(1) Progestin-treated and AGS girls and their mothers reported a higher frequency of tomboy behavior (defined as "a high level of physical energy expenditure, especially in the vigorous outdoor play, games, and sports commonly considered the prerogative of boys");

(2) Fetally androgenized girls seemed to prefer toy cars and guns to dolls;

(3) Both AGS and progestin-exposed girls attached greater importance to career plans than to marriage.

From these results Money and Ehrhardt concluded that "the most likely hypothesis to explain the various features of tomboyism in fetally masculinized genetic females is that their tomboyism is a sequel to a masculinizing effect on the fetal brain." In analyzing the nature of the increased "tomboyism," however, they determined that "fighting and aggression are not primarily implicated." Instead they suggest that dominance assertion, which manifests itself as "competitive energy expenditure," is what accounts for the greater "tomboyism" in androgen-exposed girls.[27]

A more recent study by Ehrhardt and Baker of a different group of children attempts to improve on the earlier experimental design by using family members rather than strangers as controls.[28] In addition they spoke with AGS boys, who might have been exposed to higher than normal levels of prenatal androgen, an event Ehrhardt and Baker hypothesized might make them hypermasculine. Each of the seventeen girls and ten boys received continual cortisone replacement therapy, and the girls had been surgically altered sometime during the first three years of life. It is unlikely that most of the patients underwent total clitoridectomy, although Ehrhardt and Baker never mention the extent of the surgery other than to say that the correction made the girls "normal-looking." The older girls in this study were fully aware that they had been born with an enlarged clitoris, "which usually was accepted as a minor birth defect that had been corrected." As with the earlier investigation there was no independent assessment of behavior, the authors relying on self-reports and reports by the patients' mothers.

Their results compared closely with the previous study. Parents as well as sisters and brothers of AGS girls and the girls themselves said they had a high level of intense physical energy, but saw no difference between the AGS girls and their unaffected sisters with regard to the frequency or initiation of fighting. The AGS patients took little or no interest in dolls, a marked contrast to their sisters and mothers. (Mothers, of course, had to rely on memory of their own childhood behavior.) AGS boys also seemed to have higher levels of energy expenditure than did their unaffected brothers, though again, there was no difference in how often AGS boys picked fights.* After briefly discussing some alternative explanations for their results, Ehrhardt and Baker conclude: "We are not suggesting that sexual dimorphic play, toy and peer behavior is solely determined by prenatal and/or postnatal hormone levels. We rather

---

* Dr. Ruth Bleier offers additional and damning criticism of Ehrhardt and Baker's statistical manipulation of control groups in this study.[29]

suggest that prenatal androgen is one of the factors contributing to the development of temperamental differences between and within the sexes."[30] In an even more recent publication, Meyer-Bahlburg and Ehrhardt acknowledge some of the criticisms that have been aimed at the work on prenatal hormones and aggression, at the same time holding firmly to "the theoretical framework underlying these studies . . . that of the biological bases of sex differences in aggression."[31]

Despite this recent tempering of their analysis and conclusions, a disingenuous flavor remains. Money, Ehrhardt, and other co-workers have published paper after paper, given symposium talk after symposium talk, describing studies aimed at factoring out which components of human behavior are biological and which result from socialization. They have been attacked on the one side by biological determinists and on the other by environmental determinists. They call themselves "interactionists," believing that the interplay between biology and environment creates personality. And yet their work seems most often to weigh down the biology side of the seesaw, and their attempts to explain their results in any other fashion remain feeble. It is not surprising, then, that others use Money's, Ehrhardt's, and Baker's work over and over again to argue that prenatal hormones cause sex differences in behavior. Given such use it becomes especially important to spell out in detail the fundamental problems from which the studies on AGS and progestin-exposed children suffer. Not to put too fine a point on it, the controls are insufficient and inappropriate, the method of data collection is inadequate, and the authors do not properly explore alternative explanations of their results. Let's look at each of these difficulties in turn.

Inadequate controls. AGS patients receive continual doses of cortisone, a potent drug. Although in adults such treatment can cause mood elevation and even hyperactivity, none of the controls to which AGS patients are compared received cortisone. In addition, AGS girls, born with genital abnormalities, undergo surgical correction at a young age. At least one group of psychologists offers strong circumstantial evidence that supposed sex differences in newborns (male/female differences in activity, wakefulness, and irritability) may result from the simple culturally inflicted procedure of circumcision.[32] The children in the Money, Ehrhardt, and Baker studies underwent surgery more drastic than circumcision, an event that could contribute to reported behavior differences.

Inadequate data collection. To find out about the behavior of AGS children Money and his colleagues interviewed them and their family members, all of whom knew that at birth there had been questions concerning the children's gender. Such knowledge could lead to nonconscious but nevertheless inaccurate assessments of the behavior of AGS patients. Observer bias (in this case the observers were the AGS children and their families) is well known in psychology, a fact illustrated by a study entitled "Baby X Revisited," in which experimenters told one group of subjects that the three-month-old baby with which they were to play was a boy, another that it was a girl, and then asked both groups to observe the child's behavior. All the observers ascribed sex-stereotyped behavior to the infants. One, for example, describing what he believed to be a girl (although it was really a boy) said, "She is friendly and female infants smile more," while another found a supposed female (actually a male) to be "more satisfied and accepting" than a male child would have been.[33] In the case of AGS children, the mere knowledge of early gender confusion could lead parents to assess inaccurately the behavior of their child.

Alternate explanations. Although Baker and Ehrhardt do not consider the possibilities that cortisone treatment or the surgery per se might have affected their results, or that the prenatal observations which form their data base might themselves be inaccurate, they do weigh the possibility that parental attitudes toward AGS children might have been affected by knowledge that the children were born with abnormal genitalia. In dismissing this idea, however, they simply report that in-depth interviews convinced them that the parents had "little persistent concern" about the birth defect, a result that on the face of it seems unlikely. Once again, the Baby X experiment shows how they miss the point. If the observers of Baby X believed it to be a boy (whether or not it actually was) they handed it a toy football far more frequently than they did a doll. In fact male observers *never* handed a football to a child they believed to be a girl. Did the parents of AGS children, knowing that their child had ambiguous genitalia, give her fewer dolls than her normal sister? There is no way of knowing, but clearly the parents' perception of their own feelings and behavior is an insufficient measure.

Baker and Ehrhardt consider only briefly the possible effects of clitoridectomy on the behavior and feelings of both AGS children and their parents. Furthermore, Money and his group have treated the subject in an extremely confusing manner. In their groundbreaking paper on gender identity Money, Hampson, and Hampson distinguish between clitoral extirpation, which involves the removal of all clitoral tissue, and clitoral amputation, which involves the reduction in size of an "overgrown" clitoris.[34] Almost all of their cases involve the latter, less drastic treatment, presumably leaving intact the ability for orgasm and sexual feeling (although they remain vague on this question). On the other hand, the three photographs of surgically corrected genitalia shown in *Man and Woman, Boy and Girl* seem to show total extirpation rather than amputation, although the quality of the photographic reproduction is too poor to tell for sure.

To inform a three-year-old girl about her prospective clitoridectomy Money and his co-workers tell her that "the doctors will make her look like all other girls." If the surgery results in genitalia that looks like those shown in Money and Ehrhardt's book, then these particular psychologists are in need of an anatomy lesson! They also offer to preoperative girls that classic bit of information that while boys have a penis, girls have a vagina. Yet they know full well that these are *not* analogous organs (the vaginal lip/scrotum and clitoris/penis are developmental analogues). All this, they suggest, will provide "double insurance against *childish* fears of mutilation" (emphasis added). Even if the photographs in *Man and Woman, Boy and Girl* are exceptional, there is no clear description of the appearance of AGS genitalia following surgery, and it is hard to believe the claim that AGS girls have totally normal genitalia (that is, indistinguishable from an unoperated child), although this is not to dispute the claim that the genitalia function well. The subtle effects of genital surgery on behavior, and even the likelihood of mutilation fears, cannot be lightly dismissed. Yet nowhere are these possible effects adequately discussed as contributors to the observed differences in the behavior of AGS girls.

Further work on prenatal hormone exposure. If girls exposed in utero to testosterone act like tomboys, then might not boys exposed in utero to "female" hormones behave like sissies?* A number of scientists have studied hormone-exposed children who have *normal* genitalia in order to answer this question. In these investigations the problems of genital surgery and confusion over gender designation do not crop up, and thus the hope for a more

* It is characteristic of the word imbalance in this field that the terms *sissy* and *tomboy* are not really considered parallel, and that the scientific literature never uses the former word, while often employing the latter.

clear-cut answer exists. Psychologist June Reinisch looked at a group of eight boys and seventeen girls whose mothers, in order to combat one or another complication of pregnancy, had taken a variety of different synthetic progesterones. Not all of these have the same effects.[35] When given in high doses, for example, 19-NET (short for 19-nor-17 alpha ethynyltestosterone) masculinizes developing female genitalia. Although progesterone is considered a female hormone, the body apparently converts 19-NET into testosterone, which in turn causes masculinization. (The doses in Reinisch's work were too low to masculinize female genitalia.) One of the other synthetic progesterones used in this study, MPA (medroxyprogesterone acetate), seems if anything to act as an estrogen.* In fact, it has been used instead of castration because it *lowers* testosterone production.[36] Thus Reinisch's premise that the children studied had all been exposed to a masculinizing hormone might be incorrect, but keeping that in mind, let's proceed.

Reinisch gave a test called the Response Hierarchy Test, designed to measure the *potential* for aggressive behavior, to progesterone-exposed children and to their unexposed siblings. The children studied six common conflictual situations and then chose one of four solutions to the conflict. The solutions are presented in the form of stick figures that act out choices of physical aggression, verbal aggression, withdrawal, and nonaggressive coping. Reinisch found that boys whose mothers took 19-NET or MPA during pregnancy chose the physical aggression response twice as often as their brothers, but their choice of verbally aggressive responses was just about as frequent as their brothers'. Girls whose mothers had taken these synthetic hormones chose the physical aggression response 1.5 times more frequently than their sisters, again with no difference in choice frequency for answers implying verbal aggression. In other words, the children of mothers who had received synthetic progesterone during pregnancy seemed to have a higher potential for physical aggression. Reinisch draws the following conclusion: "The observed influence of hormones . . . on later aggressive responses . . . suggests that differences in the frequency of aggressive behavior between males and females . . . may be related to natural variations in hormone levels prior to birth."[37]

I have been careful not to suggest that the children were exposed to increased levels of progesterone, but only that their mothers took

* Meyer-Bahlburg and Ehrhardt study MPA as a potential feminizing agent (see note 31).

progesterone. Here's why. By the end of the first month of pregnancy a woman's average total production of progesterone reaches 75 milligrams a day and continues at that rate during the first four to five months of pregnancy. Then production shoots steadily upward, reaching a level of 300 milligrams per day.[38] The average daily dose of progesterone taken by mothers of the children tested in the Reinisch study was 25 milligrams, an amount ranging from one-third to one-tenth of that already naturally produced by the mother. Because the actual levels of progesterone in the blood before and during hormone treatment were not measured, it is anybody's guess whether the treatment significantly augmented the naturally high levels of progesterone in the blood. The progesterone taken by pill or injection may have been nothing more than a drop in an already filled bucket.

If the progesterone treatment was physiologically insignificant, how might one deal with Reinisch's results? Sex hormones such as progesterone are given to women with so-called high-risk pregnancies, those endangered by miscarriage, premature birth, or congenital abnormalities. Often women who receive hormone treatments suffer from specific problems such as diabetes or toxemia. It seems likely, then, that all the women whose children Reinisch tested suffered more than the usual stress of pregnancy. They either feared miscarriage or the possibility of giving birth to a malformed child, or had a chronic disease such as diabetes, which causes special problems both during and after pregnancy. One recent study found that the female infants of mothers stressed during pregnancy did not feed and play as normally as did the male children in the study, and that the mothers themselves had poorer caretaking skills than did control mothers.[39] Another study explored the possible neurochemical basis for the postnatal development of aggressive behavior in children whose parents experienced unusual stress during pregnancy and following parturition.[40] Both studies are more suggestive than definitive, but they provide a hypothesis which is equally as plausible as the "male hormone causes aggression" explanation of what are, at any rate, very weak scientific findings.

Four other research groups have looked at the effects of prenatal progesterone on behavior.[41] Both their starting hypotheses and their results are the opposite of Reinisch's. For these investigations, the authors postulated that prenatal treatment with progesterone, a female hormone, might feminize both boys and girls. The first study, unpublished but referred to in a number of review articles, reported interviews with adolescents whose mothers had taken progesterone during pregnancy. Both boys and girls were perceived to engage in less physical activity during childhood, with affected girls supposedly showing even less "tomboy" behavior than controls. A different group of psychologists, using personality questionnaires, found no feminizing effect on either boys or girls born to progesterone-exposed mothers. In a third case interviews of children of mothers treated with MPA (the substance Reinisch studied as a masculinizing agent) suggested that while males from hormone-exposed mothers behaved no differently from controls, fewer hormone-exposed girls labeled themselves as tomboys. Finally, one last study reported that boys whose mothers had received an estrogen during pregnancy seemed less aggressive and athletic at certain ages. These authors are careful to point out, however, that the boys' mothers suffered from a chronic illness severe enough to require daily attention; there is no way to know how the resultant tension within the family might affect such a child's behavior.[42]

Where then, does all of this leave us? Table 5.2 summarizes the most frequently cited studies on the role of prenatal hormones in the development of aggressive behavior in humans. There are six: two study androgen-exposed children with masculinized genitalia (later corrected surgically); three examine children exposed to progesterone but born with normal genitalia; and one looks at boys whose mothers received estrogen treatment during pregnancy. It is clear from the preceding discussion that not a single one of these studies is unequivocal. Several contradict each other, while the number of uncontrolled variables makes the others impossible to interpret. The claim that clear-cut evidence exists to show that fetal hormones make boys more active, aggressive, or athletic than girls is little more than fancy, although harmless it is not.

Monkey Business

If there is no clear-cut evidence to show that different testosterone levels in adult men and women result in differences in aggression, and if no reasonable proof exists that prenatal hormones cause boys and girls to behave differently, then where is the proof that hormones make men more aggressive than women? Writer after writer has handled this question by suggesting that however tenuous the results, the data from humans point in the same direction as the more solid information obtained from other primates.

TABLE 5.2
Studies on the Effects of Prenatal Hormones on Aggressive Behavior in Humans

Subjects	Method	Findings	Critique
AGS girls	Interviews with parents and children	Greater activity, intensity, tomboy behavior[a]	No direct "blind" observations; effect of genital surgery not considered; effect of cortisone therapy not considered
AGS girls and AGS boys	Interviews with parents and children	Greater activity, intensity, tomboy behavior[b]	No direct "blind" observations; effect of genital surgery not considered; effect of cortisone therapy not considered
Boys and girls from progester-one-treated mothers	Test for potential aggressive behavior	Children from treated mothers score higher on aggression test[c]	No direct "blind" observations; actual exposure doses unknown; effect of stress in pregnancy not controlled for
Boys and girls from progester-one-treated mothers	Interviews	Boys and girls are less physically active[d]	No direct "blind" observations; actual exposure doses unknown; effect of stress in pregnancy not controlled for
Boys and girls from progester-one-treated mothers	Personality questionnaire	No difference from control[d]	No direct "blind" observations; actual exposure doses unknown; effect of stress in pregnancy not controlled for
Boys from estrogen-treated mothers	Questionnaires; direct observations; interviews with teachers, parents, and subjects	Boys from treated mothers less active and less athletically skilled[e]	All boys came from homes with seriously and chronically ill mothers; not true for the controls

[a] John Money and Anke Ehrhardt, *Man and Woman, Boy and Girl* (Baltimore: Johns Hopkins University Press, 1972).
[b] Anke Ehrhardt and Susan Baker, "Fetal Androgens, Human Central Nervous System Differentiation, and Behavior Sex Differences," in *Sex Differences in Behavior*, ed. R. C. Friedman, R. M. Richart, and R. L. Vande Wiele (New York: Wiley, 1974).
[c] June Reinisch, "Prenatal Exposure to Synthetic Progestins Increases Potential for Aggression in Humans," *Science* 211(1981):1171–73.
[d] Anke A. Ehrhardt and Heino F. L. Meyer-Bahlburg, "Prenatal Sex Hormones and the Developing Brain," *Annual Review of Medicine* 30(1979):417–30; and "Effects of Prenatal Sex Hormones on Gender-Related Behavior," *Science* 211(1981):1312–17.
[e] Irvin Yalom, R. Green, and N. Fisk, "Prenatal Exposure to Female Hormones—Effect on Psychosexual Development in Boys," *Archives of General Psychiatry* 28(1973):554–61.

Primate studies thus become the glue that binds together the fragments of evidence from human studies. But we shall see that this interspecific evidentiary edifice is a house of cards, with one weakly balanced side precariously maintaining the other in an unsteady but temporarily upright state.

The small number of investigations of the effect of prenatal androgens on male and female behavior differences (other than those related to the sex act) on nonhuman primates appears over and over again in the literature. Dr. Charles Phoenix and his co-workers injected pregnant rhesus monkeys with testosterone and studied the subsequent behavior of eight females born with mas-culinized genitalia (a scrotum and a small but normally formed penis).[43] They defined three forms of behavior which juvenile males display more frequently than females—rough-and-tumble play, play initiation, and threats. (These sex differences do not exist in all species of monkey, and even in the rhesus they may be only an artifact of captivity.[44]) The eight hermaphrodites exhibited these three behaviors with a frequency intermediate between control males and females, pointing to the conclusion drawn by Phoenix and co-workers that prenatal exposure to testosterone makes the behavior of juvenile XX females more malelike, not only with regard to reproductive behaviors such as mounting, but also with regard to nonreproductive play behaviors. Subsequent work on additional animals confirms these observations.[45]

Although the evidence in these studies could lead to the conclu-sions drawn, the investigators neglect the possibility that the behav-ioral sex differences between rhesus infants is caused at least in part by sex-specific behavior on the part of the mother. One study which looked at both maternal and infant behavior in the rhesus reached the following conclusion:

> One can best characterize mothers of males as "punishers," mothers of females as "protectors," male infants as "doers" and female infants as "watchers." *The mother plays a role in prompting the greater independence and activity that is typical of males.*[46] [Emphasis added]

In the Phoenix study, the mother may have treated the hermaphro-ditic females, which are born with malelike genitalia, more like males. Did a difference in maternal behavior rather than an effect of prenatal androgen on brain development cause anatomically masculinized females to behave more like males? For monkeys as well as humans it is probably impossible to sort out "pure" biological effects from those learned and reinforced through interactions with the surrounding environment. The significant effect of hormones may be to induce development of external genitalia, and the subsequent nonreproductively related sex differences in behavior may result from the subtle and not-so-subtle ways in which adults and peers (both monkeys and humans) react to knowing the type of genitalia possessed by each infant.

In addition to using the rough-and-tumble play of rhesus infants as a measure of aggression, some people believe there is a significant relationship between adult testosterone levels and the so-called dominance behavior of certain monkeys. The word *dominance* is often used loosely as a synonym for aggression. Animal behaviorists, however, do not use it so informally. In recent years the entire concept of dominance has come in for some heavy questioning.[47] Before looking at studies on the possible relationship between testosterone and dominance, it would be worthwhile to consider more carefully the concept of dominance as applied to nonhuman primates.

I am in a darkened movie theater, riveted to my seat. The gigantic, yet somehow sympathetic gorilla—King Kong—has just beaten his breast in a gesture of rage, ripped open the bars of his cage, lumbered through the city, and climbed the towering skyscraper where he forcefully yet gently reaches in through the window to recapture the woman he loves. King Kong's manfully dominant behavior is out of place, however, and tears fill my eyes as I watch the rescue and the retribution. If films about mythical monkeys form a part of our gut-level understanding of the idea of dominance, popular books such as Robert Ardrey's *African Genesis* fix that insight even more firmly in place: "Every organized society has its system of dominance. Whether it be a school of fish . . . or a herd of grazing wildebeest, there exists within that society some kind of status order in which individuals are ranked. *It is an order founded on fear*"[48] (emphasis added).

The concept of dominance in vertebrates, introduced to the scientific literature in 1913 to describe a pecking order among chickens,[49] gained immense popularity as a way to perceive social relations within a wide variety of animal groups. Oddly enough, during the seventy years that followed, a clear definition of the phenomenon eluded scientists.[50] Dominance is often defined in terms of aggressive or submissive behaviors. In the wild, the statement that Animal A dominates Animal B involves the prediction that most often Animal B will move over when Animal A approaches. Frequently, however, the experimenter actively defines the situation in which dominance occurs. If, predictably, Animal A gets the toy or the food the experimenter concludes that A is dominant. But as we shall see, the relationship, if any, between laboratory-induced dominance and group interactions in natural settings is tenuous.

Ardrey writes about dominance hierarchies as if they defined all social interactions. But different, sometimes contradictory, measures of dominance exist. For example, among captive monkeys that maintain hierarchies with regard to sexual behavior, grooming, and submissive-aggressive interactions, one monkey may dominate another with regard to grooming but may retreat from that same animal's aggressive displays. To complicate matters, even rankings based on the same measure change frequently—in one study, after fifteen months, only twenty of forty-five pairs of young rhesus monkeys had the same rank order.[51]

If social dominance in primates is neither stable nor consistent from one form of behavior to another, then perhaps one way to think about it is to consider the possible social functions of dominance. Scientists most often categorize these functions under the headings of leadership, reduction of aggression, and sexual behavior. Dr. Thelma Rowell, an animal behaviorist, argues that neither defense nor decisions about group movements (leadership) correlate with dominance. She also points out that hierarchies appear most clearly in situations fraught with high levels of stress and aggression. Many species establish hierarchical arrangements in captivity although they have none in the wild. Similarly, wild monkeys that live in overcrowded situations where food is scarce (such as city-dwelling monkeys in India) develop strong dominance relationships, although the same species living in a plentiful environment have none. Hierarchies appear to go hand-in-hand with more aggressive behaviors, especially those associated with a stressful environment.

Finally, the idea that dominant males get first dibs on ovulating females has, at best, only mixed support from evidence gathered in the field. One of the problems with this particular notion is the assumption that males choose females (à la King Kong) when, in fact, females do much of the choosing. And they do not always prefer the dominant male. One study used biochemical tests to determine paternity and concluded that there was no direct correlation between fatherhood and rank. Some low-ranking and even adolescent males fathered as many offspring as the highest ranking ones.[52]

Nor are male primates always more aggressive than female primates. In groups such as baboons and the great apes, males seem to be more pugnacious, but such differences are much less obvious among gibbons, siamangs, and titi monkeys. Body size may be one factor. Those species in which the male is more combative are also the ones in which the size difference between males and females is greatest.[53] Even where differences exist, however, they are environmentally conditioned. Baboon troops that live in open country have dominant males who make decisions about troop movements and get first choice of food and sexual partners. When the troop is attacked the adult males defend it, fighting a rear guard action while the females and young escape. Such behavior typifies baboons only in an open-country environment, believed to be especially stressful because of food scarcity, the large number of predators, and lack of protective hiding places. Forest-dwelling baboons, on the other hand, have good cover and plentiful food. For these animals, the core stable group consists of females and their offspring. There seem to be no male dominance hierarchies and infrequent aggression. If a troop is startled, the bigger, stronger males are the first up the trees to safety while the females, encumbered by offspring, follow up more slowly—no manly defense of the women and kids here. And it is the adult females who determine where the troop travels each day.[54] In sum, far from being a universal phenomenon among primates, dominance may not even be a useful way to describe social systems.

The question remains, however, whether dominance, when it does exist correlates with testosterone levels. For once the evidence is clear: in most cases dominance and testosterone do not correlate. One researcher[55] measured testosterone levels of wild populations of male baboons and found that they did not correlate with dominance in feeding behavior. Two other researchers were unable to correlate testosterone levels and dominance ranks and concluded that "the lack of correlation between testosterone and dominance rank in these Japanese macaques supports the hypothesis that the adult level of circulating androgen is less important than social stimuli for the display of masculine patterns of aggressive behavior."[56] A similar study with rhesus monkeys found that the most dominant male had the *lowest* testosterone level.[57] In all cases seasonal changes in testosterone levels occurred and these, not surprisingly, correlated with changes in sexual activity.

The usual argument is that testosterone levels determine social dominance. In perhaps the most interesting study of all, however, a group of researchers showed the reverse: the social context caused changes in testosterone levels. Male monkeys with access to sexually receptive females showed a sharp rise in serum testosterone *following* intercourse. If the same male, brimful of testosterone, encountered a group of strangers that attacked and defeated him, his testosterone level dropped sharply and remained low for some time. It rose again, however, if the male re-encountered receptive females.[58] This observation resembles those on stress in humans. The same research group, for instance, found that American soldiers about to do battle in Vietnam showed a large drop in testosterone levels.[59]

What About Rats?

The evidence that male hormones control aggression in humans and other primates ranges from weak to nonexistent. But the basic idea did not arise as some random scientific daydream. Here, finally, is where rats (and mice) fit in: for these rodents, there is fairly good scientific evidence that certain carefully defined types of male aggression correlate positively with testosterone levels. Aggression in wild populations of rats and mice is fairly rare. Only under certain carefully constructed—often expressly stressful—laboratory conditions does aggressive or fighting behavior show up.[60] Because the behavior of rats and mice cannot reasonably be extrapolated to humans, I offer only a brief summary of this work.

Psychologist Kenneth Moyer developed the following categories of mammalian aggression: intermale, fear-induced, irritable, territorial defense, maternal, and instrumental.[61] Two of these, intermale and maternal, show sex differences.* Rats and mice provide the best examples in higher mammals that intermale aggression is related to the level of testosterone in the blood. Two male rats in a cage will

---

* Similar lists of aggression-in-context may be made for primate interactions. For example, Fedigan (see note 52) considers that aggressive interactions may occur in daily dominance transactions, in protection of infants, when wanted resources are in short supply, when meeting an unfamiliar animal of the same or different species, in defending against predators, in the exploration of strange or dangerous areas, when there is overcrowding, when an animal has a painful injury, and so on. The important point is that *an aggressive interaction has meaning only in its social context.*

fight with one another (intermale aggression) if they receive mild electric shocks to their feet. Male *mice* need no shocks to induce them to fight, but become more combative if they have been kept in isolation or if the experimenter places them in cages with males that have been trained to fight. Castrated males fight less than normal males, but the effects of castration are overcome by injections of testosterone. Thus the relationship between fighting and testosterone is clear enough, but the implications for normal social interactions of behavior measured in such abnormal contexts is not. The physiological bases of the other types of aggression defined by Moyer are unknown. Furthermore, there is a great deal of variation between different species—for example, female gerbils and hamsters are just as aggressive as males. Because of this variability most ethologists studying such behaviors in nonprimate mammals are reluctant to use the behavior of one species to generalize about that of another.

Mirror, Mirror, on the Wall, Who's the Most Aggressive of Them All?

When carefully examined, the biological literature offers little support for the idea that high levels of "male" hormone cause human aggression. Nor does the idea that analogous sex differences in aggression exist in humans and in nonhuman primates hold much water. Yet these are two of four "facts" offered in the now classic treatise on sex differences written by psychologists Eleanor Maccoby and Carol Nagy Jacklin in support of their view that sex-related differences in aggression stem from biological differences between the sexes. Their book, *The Psychology of Sex Differences*,[62] reviews much of the work on sex-related differences in aggression in humans, and a summary of their findings appears in table 5.3. Depending upon the type of study, from 33 to 50 percent of the ninety research papers they reviewed found no significant sex-related differences in human aggression; from 46 to 61 percent found males to be more aggressive, while only 8 percent observed females to be more aggressive.

In contrast to Maccoby and Jacklin, who looked only at children and adolescents, a second group of researchers at the University of Wisconsin concentrated exclusively on studies of adults.[63] They, too, found that men were often but not always more aggressive than women. In studies using self-reports of hostility, men had a higher index for physical aggression, while in other studies they

TABLE 5.3
Summary of Studies on Sex-related Differences in Aggression

Type of Study*	Age of Subjects	Number and Percentage of Studies		
		No Sex Differences	Males More Aggressive	Females More Aggressive
Observational	1–5 yrs	12 (50%)	11 (46%)	1 (4%)
Experimental	1–5	6 (38%)	9 (56%)	1 (7%)
	6–10	6 (46%)	6 (46%)	1 (8%)
	10–17	0	3	0
	18–21	5 (33%)	9 (60%)	1 (7%)
Ratings, Questionnaires	1–6	3	3	0
	7–10	0	5	0
	18–21	4 (36%)	6 (55%)	1 (9%)
TOTALS	1–5	21 (45%)	23 (50%)	2 (5%)
	5–10	6 (33%)	11 (61%)	1 (6%)
	11–21	9 (35%)	15 (58%)	2 (8%)
	All ages	36 (40%)	49 (54%)	5 (6%)

SOURCE: Eleanor Maccoby and Carol Nagy Jacklin, *The Psychology of Sex Differences* (Stanford, Calif.: Stanford University Press, 1974).
* Maccoby and Jacklin looked at three sorts of studies: observational, experimental, and self-report. The first involved direct observation in unstructured play situations, the second was experimental, in which individuals reacted to some prearranged situation; and the third made use of self-ratings or ratings by others, such as teachers or parents.

found that the sex of the investigator, the sex of hypothetical victims, empathy with the hypothetical victims, and guilt feelings all affected findings of sex-related differences in aggression. In experiments devised to equalize these factors, female subjects acted just as aggressively as men.

Both Maccoby and Jacklin and the University of Wisconsin workers found that when sex differences appeared the male was usually more aggressive. Nevertheless, the two research groups reached rather different conclusions. The Wisconsinites felt they had no information pertinent to the possibility of a biological basis of aggression, but instead provided background information needed to design well-controlled experiments aimed at clarifying the social contexts in which men and women exhibit different aggressive behaviors. In contrast to such a context-dependent interpretation, Maccoby and Jacklin concluded that "the evidence for greater male aggression is unequivocal" and that "the male is, *for biological reasons,* in a greater state of readiness to learn and display aggressive behavior"[64] (emphasis added).

One issue in comparing these two studies is methodological: How does one explain the large number of studies that find no sex-related differences? A scientist commonly assumes that failure to show a difference (negative data) is the equivalent of lack of information. Because the studies showing differences almost always show the male to be more aggressive, Maccoby and Jacklin have followed a well-traveled path of scientific practice by ignoring the negative data and using only the positive findings to build their hypothesis. The Wisconsin group emphasizes a situational approach, trying to figure out under what conditions males will act more aggressively than females. Knowing more about when there are sex-related differences and when there are not provides a picture of how individuals behave in social situations. The negative data, ignored by Maccoby and Jacklin, tells us about those circumstances in which men and women are equally aggressive. Such an approach leads away from a simplistic view of human behavior ("males are the more aggressive sex") to a more complex one which assumes different behaviors in different social situations.

Discussions of sex-related differences in behavior almost always begin with Maccoby and Jacklin's conclusions. In arguing that sex-related differences in human aggression have a biological basis, Maccoby and Jacklin rely on four lines of evidence, none of which they themselves have analyzed in depth: (1) aggression is related to levels of sex hormones; (2) there are similar sex differences in human and nonhuman primates; (3) sex differences are found early in life before adult socialization pressures could cause them; and (4) males are more aggressive than females in all human societies for which data exist. We have already seen how weak the first two lines of evidence are. What about the last two?

Socialization and Cross-Cultural Studies

The Maccoby and Jacklin contention rests on studies of children under five years of age. Half of the investigations (see table 5.3) showed no sex-related differences, and there are strong reasons, both statistical and methodological, to question the validity of those that did show differences.[65] The difficulty revolves around the problem of observer bias. In addition to the "Baby X Revisited" study (discussed earlier in this chapter), another by Drs. John and Sandra Condry, entitled "Sex Differences: A Study of the Eye of the Beholder," relates to observations of sex differences in aggression in small children. In this case both male and female college students observed a videotape of a baby reacting to different stimuli. Half of the observers thought the baby was a girl and half thought it was a boy. The key sequence is the baby's reaction to a jack-in-the-box. On the first presentation he/she appears startled. The second time jack jumps out of the box he/she becomes quite agitated, and begins to cry even before the box opens for the third time. Those students who thought the baby to be a boy described the baby's tears and screams as anger. In contrast, those who thought the baby to be a girl said "she" responded fearfully. In other words, the behavior, crying, took on the emotional significance of anger or fear depending only upon the observer's belief about the baby's sex. The authors of "Eye of the Beholder" ask a still-valid question: "In light of these findings [would] . . . differences in observed aggressiveness in male and female nursery school children . . . hold up when the subjects being observed were cross labeled as to their sex?"[66]

Since the publication of the Condrys' article in 1976 few, if any, "blind" studies (studies in which the experimenter did not know the sex of the children under observation) on sex-related differences in aggression have appeared. Certainly none of those quoted by Maccoby and Jacklin were blind, and a serious question exists about whether there really are sex-related differences in aggression in young children. Even if there are, however, it is hard to fathom how two psychologists could claim socialization might not account for the behavior of nursery-school children. Strangely enough, Maccoby and Jacklin make no connection between their conclusion in one chapter that boys are more aggressive and another in the following chapter that parents mete out more physical punishment to boys, stimulate gross motor behavior in male infants more often than in females, and are more concerned with the development of appropriate sex typing (play attitudes and general behavior) in boys than in girls. Such differences in socialization could certainly affect the development of sex-related differences in aggressive behavior.

Maccoby and Jacklin's last argument hinges on the idea that greater male aggressiveness is a cross-cultural human universal. The observation that adult males are more aggressive in many—but not all—human cultures offers little information about a biological basis for such differences. To begin with, the anthropologists who provide the data base from which Maccoby and Jacklin work do not themselves make such generalizations about adults in different cultures.[67] Rather they describe behavior within particular cultures in *the social context in which it occurs.* To remove the behavior from that context in order to universalize it does violence to a significant component of anthropological thought and method. Furthermore, universals gleaned from adult behavior tell us little about behavior in children. If there were universal behavioral sex differences of this sort, they could as easily result from universally common socialization processes as from some biological cause.

Studies published between 1966 and 1973 identified sex differences in aggression among the children of at least eight different cultures, but because boys and girls are raised differently in each of these cultures, culture and biology remain confounded. To get around this problem, anthropologist Carol Ember studied a community in Kenya which regularly assigned work to both boys and girls. Although the villagers distinguish between "masculine" and "feminine" tasks, when there are no girls available some boys must do feminine work such as child care, housework, and fetching and gathering firewood and water. Depending upon the makeup of the particular family some boys end up doing more of this work than others. In general boys in this community are significantly more aggressive than girls, but Ember hypothesized that when boys do a fair amount of feminine work they develop a more typically "female" personality. When she tested her hypothesis, Ember found that boys who did small amounts of feminine work were less aggressive than the overall male average, while those who performed a large number of feminine tasks exhibited a 60 percent reduction in the frequency of aggressive behavior. Put another way, boys who performed few feminine tasks were three times more aggressive than girls, while those who performed work similar to girls' were only twice as aggressive. Ember also found that the difference was most striking in boys who worked inside the home rather than in public.[68]

Ember found that the different roles assigned to boys and girls affected the development of aggressive behavior. Rather than looking for a "cause" for sex-related behavioral differences, she attempted to tease apart several components involved in the development of behavior. Her work is less important for its answers than for its approach. The question of how aggression develops is complex; there are many causes, not one, and they all interact. My guess is that biology's biggest role will turn out to be highly indirect—following what one might call the "pink-blanket route." The key biological fact is that boys and girls have different genitalia, and it is this biological difference that leads adults to interact differently with different babies whom we conveniently color-code in pink or blue to make it unnecessary to go peering into their diapers for information about gender. Differences in adult behavior will elicit different reactions from children. For each individual born into this world a set of action-reaction-interactions becomes established. Many factors other than sex play a role including, no doubt, biological differences in reactivity, social circumstances of the family and the society, birth order within the family, and so on ad infinitum. For some forms of behavior, small average differences resulting from sex-related systems may become measurable. Whether this is the case for aggression is still far from clear.

An Evolutionist's Tale

A short summary of this chapter would hold that Maccoby and Jacklin's "conclusions on aggression are not supported by the data we now possess."[69] But the matter does not stop there, for there exists a different mode of analyzing aggression, derived from the study of human evolution and exemplified by Dr. David Barash's views (quoted earlier). Paleontologists writing in the 1950s offered a bloody picture of our human ancestors: "Carnivorous creatures that seized living quarries by violence, battered them to death, tore apart their broken bodies, disembodied them limb from limb, slaking their ravenous thirst with the hot blood of victims and greedily devouring livid writhing flesh."[70] This is the now familiar tale of man-the-hunter, who developed aggression, cunning, and skill to bring meat home to the hearth where women waited passively, dependent upon the competence and fearlessness of their men.[71] The sex differences in aggression emerged, so the story goes, early in the course of human evolution, thus explaining their supposed universality and confirming the contention that such differences must have a biological origin.

Since man-the-hunter was first written into the paleontological scene, there have been many in-depth critiques of the concept. Most of them take into account the facts that (1) the theory barely mentions women, (2) our early ancestors were more likely to have

153

been gatherers and scavengers than hunters, and (3) all of the evolutionary surmises about early *Homo* are based on fragmentary evidence.[72] Not all of these critiques, however, confront head-on the essential intellectual problems raised by the use of evolutionary arguments to account for human behavior. Yet these arguments have become so widespread and so much a part of popular thought (see cartoon, p. 155) that it is easy to lose sight of the fact that here too enormous controversy rages. In particular the debate embroils scientists who propose sociobiological explanations of human behavior and social organization with those who believe that sociobiology is in error scientifically and politically. The progression from Maccoby and Jacklin's allegation that in all human cultures men are more aggressive than women, to the multiply-told-tales of human evolution leads logically to an examination of the use of the Darwinian theory of evolution to explain sex differences in behavior.

Doonesbury

Copyright © 1980, G. B. Trudeau. Reprinted with permission of Universal Press Syndicate. All rights reserved.

6

PUTTING WOMAN IN HER (EVOLUTIONARY) PLACE

> The willy-nilly disposition of the female is as apparent in the butterfly as in the man, and must have been continually favored from the earliest stages of animal evolution. . . . Coyness and caprice have in consequence become a heritage of the sex.
> —SIR FRANCIS GALTON (Circa 1887)

> It pays males to be aggressive, hasty, fickle and undiscriminating. In theory it is more profitable for females to be coy, to hold back until they can identify the male with the best genes. . . . Human beings obey this biological principle faithfully.
> —EDWARD O. WILSON, 1978

Sociobiologists: Who Are They and What Do They Say?

Imagine a look into the future. The headlines leap off the front pages of newspapers across the country. ADMITTED RAPIST FREED AS JURY BUYS BIOLOGICAL DEFENSE! A feature article says the following:

> Admitted rapist Joe Smith was released today after a jury—in a landmark decision—bought the defense that sexual assault is biologically natural, and that some men—including Smith—have especially strong urges to rape. Since courts have not established procedures for confining "involuntary rapists" Smith was freed.
> There are precedents for this decision. For some years now, women committing violent acts during their premenstruum have been

156

absolved of legal responsibility after testimony that they suffered extremely from the Premenstrual Syndrome, a hormonal imbalance resulting in temporary insanity. In some courts convicted rapists have been offered the option of freedom conditional upon taking the female hormone D.E.S. Before freeing Smith the jury sifted through several volumes of highly technical testimony given by expert witnesses, all scientists trained in the field of sociobiology. According to these biologists, three different theories—the "concealed ovulation" theory, the "unsuccessful competitor" theory, and the theory of "competition between the sexes"—all lead to similar conclusions.

Expert witness A attested that "Rape is common among birds and bees and is epidemic among mallard ducks. . . . When mallards pair up for breeding there often remain a number of unmated males. . . . These bachelors have been excluded from normal reproduction and so they engage in what is apparently the next best strategy: raping someone else's female. . . . Rape in humans is by no means so simple. . . . Nevertheless mallard rape . . . may have a degree of relevance to human behavior. Perhaps human rapists, in their own criminally misguided way, are doing the best they can to maximize their fitness."[1] "In human evolutionary history, larger males were favored because of the increased likelihood of successful rape if they failed to compete successfully for parental resources."[2]

Expert witness B offered a somewhat different viewpoint. Since unlike other apes and monkeys, human females do not have a visible estrus, that is they do not advertise the fact that they are about to ovulate, human males who want to increase their own reproductive fitness—in lay terms—to pass on their own genes, face special problems. "As females evolved to deny males the opportunity to compete at ovulation time, copulation with unwilling females became a feasible strategy for achieving reproduction."[3]

Finally, expert witness C testified that when it comes to reproduction, males and females have completely different interests, their sexual patterns having evolved in totally different directions. "With respect to sexuality there is a female human nature and a male human nature and these natures are extraordinarily different. . . . The evidence does appear to support the views—which are ultimately explicable by evolutionary theory—that human males tend to desire no-cost, impersonal copulations, that there is nothing natural about the Golden Rule, and hence that there is a possibility of rape wherever rape entails little or no risk."

When asked by the prosecuting attorney whether such sexual behavior could be controlled, witness C replied, "socialization toward a gentler, more humane sexuality entails inhibition of impulses . . . [that] are part of human nature because they proved adaptive over millions of years. . . . Given sufficient control over rearing conditions, no doubt males could be produced who would want only the kinds of sexual interactions that women want; but such rearing conditions," he warned, "might well entail a cure worse than the disease."[4]

The Joe Smith case is both fact and fiction. Although no man has yet beaten a rape rap by arguing that he carries "rapist genes," some have received light sentences after agreeing to take female sex hormones, and some women have escaped criminal prosecution altogether by claiming to be victims of PMS. And although all of the writers cited as "expert witnesses" explicitly condemn rape in contemporary societies, the quotes constituting their "testimony" nevertheless come verbatim from the writings—in both popular and scientific literature—of four different sociobiologists. The words of Dr. David Barash and Dr. Randy Thornhill form the composite testimony of witness A. Barash has published one scientific paper on rape in mallard ducks and expanded his views in a popular book entitled *The Whisperings Within*, while Thornhill has written several scientific articles on rape in the scorpionfly, an insect so named because its long curved abdomen resembles the tail of the scorpion. The quotations placed in the mouths of witnesses B and C come from two other well-known sociobiologists, Dr. Richard Alexander, a professor at the University of Michigan, and Dr. Donald Symons, an anthropologist who put forth his ideas about rape in a book called *The Evolution of Human Sexuality*.

Rape is but one among many human behaviors for which sociobiologists try to provide explanation. Harvard biologist Edward O. Wilson, whose weighty book, *Sociobiology: The New Synthesis*, initiated a rush of writing, arguing, hypotheses, charges, and countercharges, begins simply. "Sociobiology is defined as the systematic study of the biological basis of all social behavior." Although apparently uncomplicated, Wilson's proposal is far from modest. Among the tasks he sets for the field are the reformulation of the disciplines of anthropology, psychology, and sociology in terms of biology to—as Wilson says—"biologicize" them. He suggests also that sociobiology will shed light on such matters as the origins of war and the development of the state and class conflict, as well as providing the basis for a natural system of human ethics.[5] Robert Trivers, professor of biology at the University of California at Santa Cruz, believes that sociobiology can tell us why parents and children fight with each other and help us understand the relationships between the sexes.[6] Richard Alexander applies sociobiological reasoning to the evolution of Western systems of legal justice,[7] while some anthropologists have joined these biologists in trying to explain the evolution of kinship systems,[8] the existence of female infanticide, the presence of social stratification,[9] and the possible biological origins of sex discrimination.[10] In short, in their enthusiasm for the vistas they believe were opened by what Wilson has named "the New Synthesis," sociobiologists seem willing to apply their approach to just about anything. As David Barash puts it, "we can safely go on our sociobiological way, confident that evolution has relevance to behavior—all behavior."[11]

Sociobiological theory is so fraught with difficulty that one is sometimes at a loss to know how to approach it. To begin with, it can purport to explain absolutely anything, a fact made clear by Wilson's broad vision of its future. Because it makes claims about matters as diverse as aggression, woman's place (in the home, of course), male and female sexuality, and so on, choosing which claims to analyze is in itself not an easy task. I chose to analyze in depth the sociobiological studies of rape because the myriad difficulties unearthed in this specific examination have general applicability to all sociobiological accounts of human behavior.

Using scorpionflies as his starting point, Randy Thornhill originally proposed a general theory of rape that predicts a more frequent occurrence of sexual assault in animal species in which the male offers some resource needed by the female for reproduction.[12] Under these circumstances a rapist could increase his evolutionary fitness by taking an unwilling female. In scorpionflies the male usually obtains and defends some food and attracts the female to partake of a nuptial meal, following which the two mate. Sometimes, however, a male without a food offering grabs a passing female and mates with her by holding on tightly to her wings and then positioning himself for copulation. Since the male incurs considerable risk in gathering food—sometimes even filching a dead insect from an active spiderweb—he may find it advantageous to mate without courting such danger. Not shy about drawing implications for human rape, Thornhill and Thornhill have recently written an extensive analysis of what they see as its evolutionary basis. They hypothesize that men rape when they "are unable to compete for resources and status necessary to attract and reproduce successfully with desirable mates."[13]

Barash's analysis is similar. He observed a population of mallard ducks living in the arboretum at the University of Washington. Mallards spend part of the year paired off in monogamous couples, and Barash paid special attention to bachelor males that mounted coupled females without first going through the normal courtship ritual. He further recorded the response of the females' partners, which often insisted on mating immediately after the females had been "raped" by an intruding male. Barash argued that the behaviors he saw were "consistent with the sociobiologic theorem that animals should behave in ways consistent with maximizing their inclusive fitness."[14]

Sociobiologists do strange things with language. All three of my English language dictionaries define rape as "the crime of having sexual intercourse *with a woman* against her will" (emphasis added). The definition contains two parts: rape is something done to a woman (although in common use we also recognize male–male rape), and it involves her conscious state of mind. For it to be called *rape* it must be against her will. When scientists apply the word to fruit flies, bedbugs, ducks, or monkeys, the common definition expands to include all living beings (Barash even includes fertilization in higher plants!) and the idea of will drops out. Yet the "instinct" of a female bedbug to avoid forced intercourse certainly holds nothing in common with the set of emotions experienced by a woman who has been raped. Using the word *rape* to describe animal behavior robs it of the notion of will, and when the word, so robbed, once again is applied to humans, women find their rights of consent and refusal missing. Rape becomes just one more phenomenon in the natural world, a world in which natural and scientific, rather than human, laws prevail.

When attacked for applying loaded terms such as "slavery," which describe particular human relationships, to behavior in the animal world, Wilson replied by asking whether other overlapping phrases should be expunged from the scientist's vocabulary. "Do they," he wrote, "wish also to expunge communication, dominance, monogamy and parental care from the vocabulary of zoology?"[15] After all, we are perfectly willing to say that animals are tired, hungry, or thirsty. Why not that they rape? As scientists, trained to recognize truth, we should be willing to call a spade a spade. David Barash writes:

> Some people may bridle at the notion of rape in animals but the term seems entirely appropriate when we examine what happens. Among ducks for example, pairs typically form early in the breeding season, and the two mates engage in elaborate and predictable exchanges of behavior. When this rite finally culminates in mounting both male and female are clearly in agreement. But sometimes strange males surprise a mated female and attempt to force an immediate copulation, without engaging in any of the normal courtship ritual and despite her obvi-

ous and vigorous protest. If that's not rape, it is certainly very much like it.[16]

The main difference, however, between observing that an animal is, for example, sleepy and that it rapes lies not in an accurate description of its behavior, as Barash suggests, but in the meaning of the words to human beings. The existence of sleep is for all humans a fact of life, controlling in some sense one's plans and daily activities. The existence of rape, on the other hand—or rather the fear of it— controls lives in quite another way. For example, it limits my evening activities, but it does not limit my husband's. It dictates the route I walk to work each day, but it does not dictate my male colleague's, who lives next door. It restricts what parts of the world I can travel to and with whom, but it does not restrict the male undergraduates whom I teach. It even prevents me from daydreaming too intensely when I walk about outside, lest I lose track for one fateful moment of a car slowing down as it passes or footsteps approaching from behind. Rape is not just another word. It signifies an important fear that affects my life, a fear that influences the lives of most women but few men in our culture.

Students of animal behavior face a fundamental problem—how to name the observed activities of animals. The difficulty is not trivial; we are limited by one of the very things that separate us from other animals, the development and use of spoken language. When ethologists study a particular behavior, they think long and hard about the question and come up with various compromises, usually accompanied by long qualifications of what is *not* meant by the use of a particular word. In contrast many sociobiologists— especially those using examples from the animal world to draw analogies with human behavior—seem curiously oblivious to the problem. Indeed this is one of the major grounds on which such scientists have been attacked. In using the word *rape*, Barash, Symons, Thornhill, and others have transformed its meaning. First, to describe certain animal behaviors they use a word originally applied to a human interaction, one that includes within its definition the notion of conscious will. Then they employ the animal behavior (named after the human behavior) in theories about rape in human society. In the process they confuse the meanings of two different behaviors and offer a natural justification for a human behavior that Webster's calls criminal. *This linguistic hat trick characterizes virtually all of human sociobiology.*

Until recently the use of the word *rape* to describe animal behavior occurred only rarely, but the practice became increasingly more common after 1975. This timing takes on significance when one looks at what the feminist movement has had to say about it all. By the mid 1970s several scathing books on rape were either newly published or in the offing. Women's groups around the country started rape crisis centers and counseling programs, while protests against a legal system that protected the rapist but harassed the victim began in earnest. In short, rape became a hot topic. The sudden increase in the use of the word in the biological literature, as a response to the furor raised by feminists, was at the very least a non-conscious attempt to establish rape as a widespread natural phenomenon and thus deflect and depoliticize a subject of intense and specific importance to women. The attempt to defuse the political explosiveness of rape, thus trivializing its effect on women's lives, is what provokes such spontaneous anger from the critics of sociobiology. But angry though we may be, we must still contend with Barash's argument: if scientific observations reveal rape in nature, must we not face up to this reality?

Do Animals Really Rape?

Not surprisingly for a field that has flung its net so far and so wide, there are many kinds of sociobiologists, and they do not all agree with one another. Some use an approach called game theory to speculate about possible reproductive strategies. Others make arguments based on population genetics. Some are careful scientists making well documented, detailed studies on specific animal behaviors, while remaining extremely cautious about applying their results to humans. Others are more than willing to leap from ants to chickens to baboons to humans and back again, with only the most casual glance at the intellectual chasms yawning beneath their feet. Before Wilson laid claim to sociobiology as a new field, warranting its own name and having its own unique literature and approach of study, scientists interested in evolution and animal behavior worked mainly in one of three traditions—population genetics, ecology, and animal behavior (known in the halls of academe as the field of ethology).

Randy Thornhill is an ecologist. During certain times of the year his research work takes him to the field daily, where he watches and records the behaviors of particular predatory insects from sunup to sundown.[17] When he's not actually observing the species of his choice he may be out sweeping the bushes with nets so that he can catch and classify the kinds of prey available to the flies that are the primary objects of his studies. Like any competent field biologist, he may develop theories in the field, but he uses his ingenuity to test them experimentally in the laboratory. Such is the case with his analysis of the behavior in scorpionflies which he labels rape.

As is the fashion with scientific publications, Thornhill first defines his topic and cites supporting examples from the literature. "Rape," he begins, "is forced insemination or fertilization." To substantiate his claim of rape in nature, he mentions in addition to observation of rape in ducks "behavior suggesting heterosexual rape" in fish.[18] Yet the fish to which he refers do not have internal fertilization; instead, they follow a mating ritual which serves to synchronize their behavior. Females and males release their gametes into the open sea, where the fusion of egg and sperm takes place. Sometimes an intruding male (one that has not done his courtship bit) rushes in, depositing his sperm near the eggs as they are shed into the water. It is this "stolen" fertilization—not involving copulation at all—which Thornhill finds suggestive of heterosexual rape. In this example, then, he equates fertilization with rape. He also invokes two instances of "homosexual rape." When I tried to follow up the first, a claim of homosexual rape in bedbugs, I found only a reference to an unsigned article in Newsweek, a most unusual source to cite in a scholarly publication. The second example came from a paper published in the journal Science, in which the authors assert the existence of homosexual rape in a parasitic worm and remark at length on its possible evolutionary origins, but never describe the behavior they so enthusiastically discuss.[19] As a result it is impossible to judge for oneself the propriety of the label.

In marked contrast, Thornhill clearly describes the events in his study. Most of the time male scorpionflies obtain food (usually dead insects) and secrete an attractant to entice females to feed and then copulate with the male. Sometimes, however, a male without a food offering rushes toward a passing female, attempting to grasp a wing or a leg with muscular claspers that form part of his genitalia. If he gets a grip he tries to place her in a position that would permit copulation, and he frequently succeeds. Females generally run from males without nuptial gifts, and ones grabbed in this manner fight strongly to escape. Thornhill argues that he can only label this foodless mating as "rape" if the female's escape behavior is genuine rather than "coy" resistance. To investigate he set up two situations in his laboratory. In one cage he put males and females together without any food, while in a second cage he introduced females, after giving the males a bunch of dead crickets which they each staked out, one-male–one-cricket.

Most (90 percent) of the males without food offerings tried to copulate with the females, but only 22 percent (eight out of thirty-six) succeeded, and of these only half actually transferred sperm into the female reproductive tract. The latter observation is the crucial one both for consideration of the possibility that such behavior is inherited and for a view of rape as a particular *reproductive* strategy. Thornhill then took the thirty-two females that had avoided mating in the cricketless cage and put them into the cage with the goodies, where most (twenty-seven out of thirty-two) mated almost immediately. Since he had thus shown that the same females that struggle to elude males without food offerings mate rapidly with males with food, he concluded that the females' previous avoidance behavior was not simple "coyness" and that rape had indeed occurred.

Thornhill writes, "Although successful rape [that is, including insemination] may be infrequent, it is an appropriate behavior for a male to adopt when he is aggressively excluded from possession of a dead insect."[20] In this context "appropriate behavior" means one that will increase the male's reproductive success, thus augmenting his "genetic fitness." By analogy, an "unfit" male would be one that could not get his own food cache or manage to mate any other way.

Getting a food offering in the first place is a risky business. Males have even been known to attempt to rob dead insects from spider webs, often losing their lives in the process. Forced mating, then, might increase a male's relative fitness by helping him avoid such risks. If this were true, Thornhill asks, "Why do not all males use rape as their primary behavior?" The answer: a hypothesis that females that feed on food offerings lay many more eggs than ones that do not, the trade-off occurring between the risk taken in obtaining food and the enhancement in female fertility if she gets more to eat.* Thornhill further suggests that since no food is

---

* Evolutionary biologists consider fitness to be the ability of an individual to survive and reproduce. In estimating fitness one must consider an individual's life span, how many times it reproduces, how many offspring come from each mating, and how likely those offspring are to survive to reproductive age.

forthcoming, rape decreases female fitness—a point that is difficult to understand given that females mate up to five times per week and must obtain food on at least some of those occasions. This allegedly lowered female fitness, Thornhill believes, substantiates his assertion that the phenomenon under study is rape. As is characteristic of all sociobiological treatment of male-female interaction, Thornhill emphasizes conflict:

> Female fitness is probably reduced by rape because the female gets no food from the rapist and must therefore expend time and energy in finding a male with nuptial food or in finding food on her own in an environment with abundant predators. . . . Male fitness is enhanced by rape because predation-related risks . . . can be avoided.[21]

Since these males compete heavily with each other for food, many end up with none. An empty-handed male cannot convince a female he's worth the time of day. His unfitness is plain to see. Thornhill suggests that in species in which males normally make no nuptial offerings it is harder for a female to tell how fit he is, so that males may get to mate just by doing a good job of courtship. In this case a less fit male can trick a female into mating with him because she can't be sure how fit he really is. "Thus," Thornhill argues, "selection for heterosexual rape should be stronger on males of species with male resource control than on males of other species."[22]

Thornhill does not shrink from the task of applying his work on rape in insects to the human condition. In an extensive discourse on human rape he and a co-author define rape as forced copulation which reduces the ability of the female to make her own reproductive choices. In explicating this definition they write:

> Copulation by a man with women who depend on him (e.g., a male employer copulating with his female secretary or a male slaveowner with his female slave) is not *necessarily* rape . . . by our definition because the female need not be denied the option of gaining benefits that exceed the costs to reproduction (job security or salary [secretary]; resources or higher status [slave]).

In the same paragraph they suggest that because high status men have so much to offer women who are dependent upon them, those women will rarely see copulation with their bosses or owners as maladaptive, but instead will use the interaction to further their own reproductive interests. Thus, according to Thornhill and

Thornhill, "high status men probably rarely actually rape."[23]

In the Thornhills' world view human males "compete with each other for relative status, including wealth and prestige." Resulting from such competition is a gradation of successful males:

> At the top of this continuum are the big winners. . . . These males are represented by men with multiple wives in preindustrial human societies and by the highest executives in large corporations, powerful politicians, *leading scholars,* and outstanding athletes and entertainers in many industrial societies. [Emphasis added]

On the other hand, they see a large number of men who leave the competition with limited reproductive possibilities, a small amount of money, and a struggle to hold on to just one wife and a small family. At the bottom of the heap are "the big losers: those men who are excluded from a share in the wealth, prestige and resources, and thus access to desirable mates. . . . it is those human males who have the greatest difficulty climbing the social ladder who are most likely to rape."[24]*

As with many lines of reasoning in sociobiology, the Thornhills' offers the appearance of logic and simplicity although it is in fact a highly layered and encoded argument. In order to get a solid grip on it one must peel off the various underlying assumptions, place them in the context of the theory of evolution, and see which ones have evidence to back them up and which are nothing more than guesswork. We ought also to ask what kind of evidence Thornhill would need—at least for scorpionflies—to prove his hypothesis.

The Thornhills and others[25] make a startling claim. They argue that modern evolutionary theory holds within it the lesson that rape is but one of a number of co-equal reproductive strategies to have evolved through the millenia. They feel we must recognize this fact precisely because rape in our modern cultures is so reprehensible and hope to use "applied sociobiology" to help society deal with the problem. Their argument is that males try any means to "maximize inclusive fitness"; females try to acquire as many "parental resources" as possible; males look for "low-cost matings"; the female "reproductive strategy" is to seek the best possible male. Sociobiologists generally view reproduction in terms of male-female

---

* Note that this exposition implies also that social and economic class result from competition among men with different biological endowments. Social forces seem to play only a subordinate role in the formation of class and status hierarchies.

conflicts, of which rape is simply one example. Many modern Darwinists, however, believe that this sociobiological mode of argumentation derives little support from Darwinian theory.

Evolution: The Modern Synthesis

Darwin documented the existence of variation in the natural world: for example, in one species of snail some individuals may have shells that coil to the left while others' coil to the right; in some regions of the eastern United States coal black "gray" squirrels are quite common; and most mammals have naturally occurring albino forms. Variation abounds in living organisms. Furthermore, as Darwin wrote in the *Origin of Species*, all living things "struggle" to stay alive. Many more offspring, eggs, seeds, or spores—the means of continuing the species—are produced each generation than can possibly survive.

> Can we doubt ... that individuals having any advantage, however slight, over others would have the best chance of surviving and procreating other kinds? On the other hand, we may feel sure that any variation in the least degree injurious would be rigidly destroyed. This preservation of favorable individual differences and variations and the destruction of those which are injurious, I have called Natural Selection, or the Survival of the Fittest.[26]

While Darwin knew that animals varied and that at least some types of variation were heritable, he understood nothing about the mechanisms of inheritance. Furthermore, Darwin provided no experimental demonstration of natural selection in action. Thus his ideas met with strong resistance from two rather different quarters. On the one side were people who responded to the religious challenge brought by scientific theories of evolution; the attacks on Darwin from the clergy were strong indeed. But by the turn of the century the era of modern genetics picked up steam, and the theory of natural selection experienced attack as well from the scientific community on the grounds of insufficient evidence. In fact it was not until seventy-five years after the first publication of the *Origin* that biologists reached some consensus about the general mechanisms of evolution through natural selection. The book responsible for bringing this unity, however temporary, was Julian Huxley's *Evolution: The Modern Synthesis;*[27] Wilson's title, *Sociobiology: The New Synthesis,* is a conscious variant on Huxley's, a title familiar to most biologists.

The period between the *Origin* and *The Modern Synthesis* was a busy one. To begin with the work of an obscure Czech monk named Gregor Mendel came to light. During the first decade of the twentieth century his research became not only widely known, but was a stimulus for an explosion of discoveries in the field of genetics. Scientists devised rules to describe the passage of genetic information from one generation to the next. They began to work out the pathways leading from a gene to some measurable trait, finding that the connections were not always direct. They began to devise mathematical accounts of how specific genes might maintain themselves in a population. And they discovered genetic mutation. Geneticists, so absorbed in their rapidly developing field, on the whole ignored Darwin's powerful use of data from natural history (about which this new breed of scientist knew relatively little). Instead they proposed that the mechanism of evolution involved something called *mutationism.* Natural selection might act as a negative pressure eliminating harmful mutations, but positive, creative change must, they held, occur through the continuous pressure of new, small, helpful mutations.[28]

At the same time that the Mendelians, working mostly with fruit flies, mice, and corn, established the particulate nature of inheritance of certain traits (Mendel's peas were either wrinkled or smooth, not intermediate), other geneticists realized that not everything behaved in such convenient fashion. Traits such as height result from the interactions of a number of different gene combinations with one another as well as with the environment. Humans are not either six feet tall or five feet tall. They form a continuum of variation from midgets to giants. Geneticists coming from an intellectual tradition somewhat different from that of the Mendelians termed such continually varying traits "quantitative" and developed mathematical methods of analyzing their inheritance.

Sociobiologists believe that most human behavior is genetically inherited. To the extent that this is correct, such behavior must be treated as a quantitative trait, since many genetic factors are certainly involved. Except when using metaphoric hyperbole no sociobiologist seriously argues that single genes determine complex human behaviors. The quantitative geneticists also used a statistical approach to the study of evolution. Instead of looking at the result of a mating between two individuals (for instance, a black and a gray squirrel), they asked questions about the distribution of genes in entire populations, developing concepts such as "gene pools" to describe the total genetic resources of a group of animals or plants which breed with one another. Looking at the collective genetic traits of a population is a key methodology of mainstream evolutionary theory. It grew out of the analysis of quantitative traits. Sociobiologists, however, write about behaviors as if they were particulate traits, even though they know this is an oversimplification. They end up, therefore, using evolutionary theory based on the inheritance of particulate traits to analyze behaviors that—if they *have* a genetic basis—must be multiply determined. In short, they inappropriately mix their analytical media.

During the first forty years of the twentieth century Mendelians and quantitative geneticists alternately quarreled with or ignored each other. Finally, out of the strife born of different traditions of knowledge came the Modern Synthesis so thoroughly considered in Huxley's book. It reformulated Darwin's proposals in up-to-date terms: individuals within most sexually reproducing populations vary to some degree in their genetic makeup. New genetic variants arise by mutation; in addition variability increases because the chromosomes in which genes reside constantly undergo rearrangements. Gene A may have originally lain next to Gene b—and may still in some members of a population—but since homologous chromosomes interchange parts in each generation through a process called *crossing over*, in other individuals Gene A may lie next to Gene b, and Genes A and b may interact when they lie together in new, even unpredictable ways. Because during sexual reproduction chromosomes rearrange and change their associations, the initial variation introduced by mutation increases greatly.

Sociobiologists frequently write about evolution as if it were a single, uniformly agreed-upon theory. In fact, many currents, sometimes conflicting, contribute to the stream of evolutionary thought. Although the Modern Synthesis brought Mendelians and quantitative geneticists into general agreement, even today their intellectual descendants bring very different viewpoints to the study of evolution in populations.[29] Arguments continue to rage over the degree to which genetic variation is an ever-present feature of populations, and whether evolution proceeds through the effects of many genes

169

with minor effects working in concert rather than through the appearance at auspicious moments in the life of a population of rare genes with major effects. There is not yet *an* accepted analysis of evolution but a number of traditions of analysis, some predominating in the United States, others in Europe.[30] If one were to read only the sociobiological literature, one would lose sight of the intellectual richness of the study of evolution—missing would be some of the difficult and fascinating intellectual struggles in which students of evolution find themselves.[31]

Sociobiologists focus on individuals' attempts to *maximize* their genetic fitness. The Modern Synthesis, however, contains no arguments about perfection. Organisms do not strive to maximize their ability to fit the environment in which they find themselves; in fact, over the long haul such maximization could spell extinction. The more perfectly adapted an organism is to its environment, the less flexibility it has should the environment change. A versatile animal such as a raccoon has managed rather well, even in the heart of large cities, to live off of human refuse; it remains annoyingly successful despite the constant diminishment of its original forest environment. In sharp contrast, the California condor, shy and environmentally limited, is more likely to end up like the passenger pigeon, stuffed into a museum display cabinet.

The Indian and African rhinoceri provide another illustration of how the notion of maximization of fitness deviates from mainstream evolutionary theory. Both use horns for defense, but the Indian rhino has only one while the African sports two. Is there some reason that two horns are more adaptive in Africa while one horn promotes better survival in India? Probably not. It is more likely that one-horned, two-horned, and hornless varieties of rhinoceros existed on both continents, but that in India by chance the population experiencing selective pressure had a higher frequency of the genes for one horn, while the ancestral African population happened to have had a higher starting frequency of genes for two horns. In each case there may have been the same degree of selective pressure favoring horned over hornless varieties. But in the Indian population, which already had a higher frequency of single-horned genes, the pressure selected for one horn; while in the African rhinos the quickest route to success was through the increase in frequency of two-horned individuals (see figure 6.1). In other words, there is more than one way to horn a rhino.[32] In fact in most populations of any size and complexity there are multiple ways to

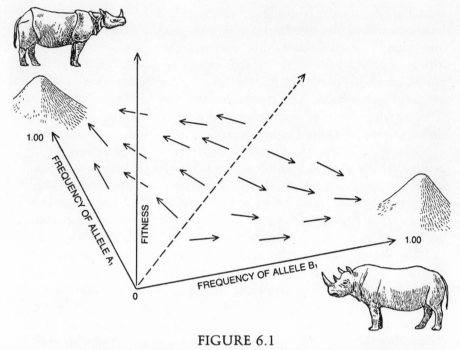

FIGURE 6.1

Alternate Evolutionary Paths of One- and Two-Horned Rhinoceri

Alternative evolutionary paths may be taken by two species under similar selection pressures. The Indian rhinoceros has one horn and the African rhinoceros has two horns. The horns are adaptations for protection in both cases, but the number of horns does not necessarily constitute a specifically adaptive difference. There are simply two adaptive peaks in a field of gene frequencies, or two solutions to the same problem; some variation in the initial conditions led two rhinoceros populations to respond to similar pressures in different ways. For each of two hypothetical genes there are two alleles: A_1 and A_2, B_1 and B_2. A population of genotype A_1B_2 has one horn and a population of genotype A_2B_1 has two horns.

NOTE: Richard Lewontin, "Adaptation," *Scientific American* 239(1978):225. Copyright © 1978 by Scientific American, Inc. All rights reserved.

skin the selective cat and the one that is chosen is not always the best; it is the most quickly attainable.[33]

Sociobiologists emphasize natural selection to the exclusion of other mechanisms of evolution. Evolution can occur, however, without the intervention of either adaptation or natural selection. Among the best-studied examples are colonizations of previously uninhabited islands. Imagine an interbreeding population of birds living on the mainland. Within that population might exist genetic variations in feather color, for example a solid blue variety and a speckled pattern. Suppose that by chance a small number of the speckled birds got blown onto an island where no birds of that species previously lived. Should the birds survive, breed, and inhabit the island, the entire island population would be speckled rather than blue. A biologist who subsequently studied these two geographic locations would find two bird populations with different gene pools, one including two feather color variants, the other only one. Evolution—defined as a stable change in the frequency of genes in a population—has occurred. But the cause was a chance natural event, not natural selection. And the biologist studying the population—if he or she knew how it originated—would certainly not argue that speckled birds are better adapted to island life.

Island colonization illustrates but one of the ways in which evolution may proceed without either adaptation or natural selection, but there are many others. Consider albino animals of the arctic. Hungry predators may have unwittingly selected for the albino color variant by eating more of the easily visible dark-colored variants. But albinism, which affects coat color, also causes severe problems with eyesight; albino animals generally cannot see as well as their colored counterparts. It is hard to imagine an adaptive "use" for poor eyesight. For arctic-dwelling mammals it just comes along as part of a package deal. Probably all genes are (like the gene for albinism) *pleiotropic*, that is, they have more than one effect (remember the example of sickle-cell anemia from chapter 3). Many traits common in populations of plants and animals have undoubtedly, as has nearsightedness in albinos, come along for the ride. Such traits have evolved, but they have not been selected for, nor are they adaptive.

In addition to evolution without natural selection, a population can also experience selection without adaptation, and adaptation without natural selection. A gene that doubles an individual organism's fertility would spread rapidly in a population. But if the population does not increase its food supply some of the extra animals produced will starve and the number of individuals who survive at any one time will remain more or less steady. While the population can in no way be said to have become better adapted, natural selection favoring individuals carrying the genes for higher fertility has certainly occurred.[34] Or, consider a species of shellfish which uses sand and mud grains to make its shell. By using local construction material, the animal ends up matching its background quite nicely, and the color coordination may well provide good camouflage. The well-suited shell color is an adaptation, but it has not occurred by natural selection. Instead it is a by-product of the fact that this particular shellfish uses the ground it sits on to build its shell.

PUTTING WOMAN IN HER (EVOLUTIONARY) PLACE

The lesson from all of this is that one must proceed with caution when trying to decide whether the evolution of a particular trait has occurred by natural selection. It can even be difficult to know whether one has correctly identified a legitimate trait. Writings on the evolution of the human hand illustrate this particular dilemma. What was selected for? The whole hand, the opposable thumb, the flexibility of the fingers? We can never be sure. The problem arises because scientists are trained to break things down into component parts. The hand, however, functions as a whole unit, and losing sight of this can lead one into hopeless philosophical quagmires. The disagreements between evolutionary biologists who work by progressive atomization of traits and ones who believe most such traits to be artificial are profound and probably without resolution.[35]

The notion that evolution is best studied by identifying traits and then trying to figure out how they might help an organism better adapt to its environment predominates today in American scientific thought. But merely making up a plausible account will not do—good scientific inquiry requires careful consideration of nonadaptationist explanations, such as those just exemplified. Huxley understood this quite clearly, despite the fact that his book especially emphasized the processes of natural selection and adaptation. In its early pages Huxley explains the elements of proof in the study of evolution, emphasizing clearly that biology is first and foremost an experimental science:

> A particular shade of human complexion may be due to genetic constitution for fair complexion plus considerable exposure to the sun, or to a genetically dark complexion plus very little tanning; and similarly for stature, intelligence, and many other characters. *The important fact is that only experiment can decide between the two.*[36] [Emphasis added]

The phenomenon of industrial melanism comes about as close as one could hope to a fully proven case of evolution through natural selection. Living in the industrial regions of northern Europe is a moth called *Biston betularia*, commonly referred to as the peppered moth because of its light mottled appearance, a color pattern that blends in perfectly with the lichen-covered bark of trees in the area. As more and more industry developed in Europe, large areas became polluted with huge quantities of smoke particles, measurable in tons per square mile per month. As the particles darkened surfaces they (along with sulfur dioxide gas) killed the tree-growing lichen, exposing the dark underlying bark.[37] No record existed of a dark form of the peppered moth until the mid 1800s, when dark-colored variants appeared with some frequency and, in the most heavily polluted regions, became more common than the light form for which the moth had been named.

Geneticists interested in this moth established that the different colors were inherited. Following its initial appearance (presumably a spontaneous mutation), the first dark-colored moth spread much more rapidly in the population than would be expected if the dark moth and its descendants survived neither more nor less well than the lightly peppered variety. Scientists used ecological studies to decipher this puzzle. By watching the moths during the daytime when they are inactive and rest on tree trunks, it immediately became clear that the dark form was very difficult to spot in industrial areas where the bare, darkened bark served as the background, while in unpolluted regions the peppered variety blended in almost perfectly with the lichen-covered bark. Furthermore, they saw birds swoop in and eat resting moths. In an ecological experiment, one scientist marked and then released both darkly and lightly colored moths and retrapped them later, finding that in polluted areas he could recover fewer light than dark moths. The combination of observations led him to conclude that the birds ate more of the light moths in polluted regions, while the reverse was true in nonindustrial areas.[38]

Natural selection in the persona of hungry birds enabled the darkly colored variants to spread rapidly in the moth populations of industrialized England. Or, as a population geneticist would phrase it, a change in the environment led to a change in genotype frequency (more dark than light genes) within the population. The birds acted as selective agents: in unpolluted environments the light-colored moths had a selective advantage, making them fitter than the dark moths. The light moth living in unpolluted regions had a greater likelihood of producing more offspring than the dark moth; in polluted parts of the country the reverse was true.

A method of investigation emerges from this example. The population studied could be shown to have color variations (variation measured, rather than hypothesized). Breeding experiments established the genetic basis of the variation. The changes in gene frequency were measured, and the selective agents first hypothesized, then tested by field experiments. Even so it now turns out that selection has shown up at an unexpected stage of the moth's life cycle, a finding which even in this well-studied case leaves us partially ignorant of the workings of natural selection.

If the Modern Synthesis represents such a rich and complex body of thought and sociobiology such an impoverishing oversimplification, why did Wilson's book and the others that followed receive such instant acclaim? It is of course true that the explanation and defense of the status quo contained within this body of writing offers a certain psychological and political sense of well-being, but that offers only a partial explanation of the phenomenon. The missing puzzle piece is that Huxley's Modern Synthesis did not solve all remaining questions about evolution. Sociobiology developed as one of a number of attempts to remedy the shortcomings of Huxley's account of evolution.

Evolution: The Postmodern Dissolution

For about twenty years, the Modern Synthesis worked. In fact it worked so well that—in one of those contradictions that philosophers love—its very success brought new controversies. Using the Modern Synthesis as a framework, biologists analyzed a host of natural phenomena, putting meat on the bones of evolutionary thinking. It seemed reasonable to extend the insights gained about the process of evolution to the study of social behavior. And herein lay one of the troubles: behaviors were easy to describe, but accounting for their evolution was another matter altogether. For example, in her book on the mountain gorilla, Dian Fossey vividly describes the death of Digit, a male gorilla slaughtered by poachers as he defended a gorilla band consisting of his siblings, parents, and nonrelatives.[39] He did not just beat his chest and run, saving his life while permitting the others a chance at escape. Instead, he stood his ground and battled to the death. Digit's behavior appears to have been altruistic; he died to save the lives of the other troop members. His action, though, presents a puzzle to the evolutionist. If altruistic individuals are less likely to survive than "selfish" ones, *and* if the behavior in question is genetically determined, then how could it have become common in the population? More generally stated, the question becomes one about the evolution of cooperative behavior. That such behaviors evolved is certain. *How* is far from clear. This is one of the central problems sociobiologists have set out to solve.

The first stab at an answer came with the publication of the theory of group selection. It was proposed that natural selection could act on a group as a unit, selecting characteristics that might benefit the species as a whole even if it sacrificed some individuals. I remember grappling with this idea as a student, intrigued by the problem it addressed yet puzzled by the vague formulation of the solution. Biologists now agree that the idea of group selection, as originally put together, must be wrong. But from that idea has arisen a furious debate about the level of selection. Does natural selection operate on whole species at once? On subspecies? Darwin thought it acted on individual organisms, but after Mendel's rediscovery many came to think of selection occurring for a particular constellation of genes (genotype). More recently, impassioned arguments have been offered for the idea that individual genes are the units of natural selection. Related to the disagreements over levels of selection are disputes about the definition of *fitness*. We must, then, continue to view sociobiological ideas as part of a heavily debated scientific controversy; to understand the elements of this debate we need to examine more closely the meanings of phrases such as *kin selection, inclusive fitness,** sexual selection,* and *parental investment.*

David Barash defines his view of the "Central Theorum of Sociobiology: Insofar as a behavior in question represents at least some component of the individual's genotype, then that behavior should act to maximize the inclusive fitness of the individual concerned."[40] Although sociobiologists write about the "maximization of inclusive fitness" as if it were accepted biological dogma, this new and controversial idea is one of the weakest links in the sociobiological chain of reasoning. We have already addressed some of the difficulties of the notion of maximization, and we must now consider what is meant by *inclusive fitness* and what it has to do with the "problem" of altruism.

An altruistic act lowers the actor's individual fitness (chance of survival), and that lowering ought to make it less likely that any possible genetic basis for the behavior survives in succeeding generations. Yet such behavior does exist. The solution to the problem, suggested by biologist W. D. Hamilton, lies in the related concepts of *inclusive fitness* and *kin selection.* For simplicity's sake let's imagine that the defense behavior of an adolescent male gorilla is

---

* Earlier in this chapter (p. 164) I introduced the more generally accepted concept of *individual fitness.*

encoded by a single gene, while remembering that animals within the troop are genetically interrelated. Because siblings, on average, have 50 percent of their genes in common, Digit's brothers and sisters stand a 50 percent chance of carrying the hypothetical "troop defense trait." Using greater precision than in the earlier proposal of group selection, Hamilton demonstrated mathematically that the troop defense trait could remain in a population generation after generation if the behavior increased the likelihood of sibling survival by more than two times, that is, by making it twice as likely that a family member with half the defender's genes would survive.

An analogous relationship would hold for altruistic acts affecting other family members. Nephews and nieces, for example, have on average one-fourth of their aunt's or uncle's genes. Thus, should the genes that hypothetically code for any act—for example, a warning call—even if the act led to the death of the caller, make it four times as likely for a niece or nephew to survive, the trait would remain in a population. Hamilton generalized these arguments into the concept of inclusive fitness, which holds that the effect of a genetically inherited trait can be measured by its effects on the fitness of an individual *and* on that individual's kin, depending on the degree of genetic relatedness of the kin.[41] With this approach fitness, rather than being an individual trait, becomes the property of several individuals related by family ties; selection for increased fitness occurs on an individual *and* its kin—hence the term *kin selection*. Hamilton's formulation was quite successful in explaining certain specialized problems in the behavior of social insects such as bees, ants, and termites.

Barash, however, raised Hamilton's limited concept of inclusive fitness to a central theorum. Indeed the invocation of inclusive fitness to explain the evolution of an enormous range of complex behaviors is one of the distinguishing features of sociobiology and, because it is not easily subject to experimental test, it is also one of its weakest. Barash himself recognizes the difficulty, "I know of few studies that have even measured individual fitness," he writes. "In most cases we must be content for now with measures that we *intuit* as being related to fitness . . . such as efficiency in foraging, avoiding predators, success in acquiring a mate, or often, simple copulation frequencies"[42] (emphasis added).* An argument relying on intuition rather than on something we can measure—however

* Measuring fitness is a difficulty for all of evolutionary theory. For a more detailed discussion of the problem see Appendix, pp. 223-24.

imperfectly—removes itself from the realm of conventional science, a realm in which one devises hypotheses and methods to test them, proving them "right" because they plausibly pull together most (but rarely all) available observations, or disproving them by finding they cannot handle a significant fact or observation. One standard method scientists use to shore up a position is to consider alternative hypotheses and show them wanting.

Barash, for example, uses the concept of inclusive fitness to talk about the evolution of menopause, a phenomenon that appears to be uniquely human. He wonders why women stop reproducing even though they may have as many as thirty years of life left. Since he figures that everybody ought to be out there maximizing their fitness by constant reproduction, the puzzle seems great to him. Even allowing that older women face greater risks in childbirth is insufficient.

> Although the risks ... are higher for older women ... they should still be selected, as are men, to die trying. The genes carried by older women who average just one additional child because of some Herculean, possibly suicidal reproductive effort should still surpass statistically the genes carried by other women who elected instead to enjoy a peaceful, unencumbered but unproductive old age.*

Barash answers his own puzzlement by invoking inclusive fitness:

> As grandmothers, middle-aged women can provide for their fitness by caring for their children's children instead of bearing more themselves.[43]

He shores up this position by citing the universality of grandmothering, suggesting that his view of the origin of menopause is consistent with "the calculus of evolution."

And indeed it is. But so, too, are other explanations. Menopause might, for example, have both a biological and a cultural origin. Our protohuman ancestors never lived long enough to have more than a few postreproductive years. As humans developed the capacity for culture, however, they increased their average life span by improving their diet, building better shelters, and learning to cure the ill—all changes wrought by conscious human activity. Living long enough for the body to age and to enter that part of the life span in which menopause even shows up is recent (on the geological

---

* Note how Barash equates human sexual activity with reproduction in this passage.

time scale) and is the result of human cultural activity. Menopause itself then is but a by-product of aging, which in turn became possible only through the advance of human culture.

I would argue with Barash that my explanation is more plausible than his. He, presumably, would disagree. I cannot think of any evidence either of us could bring up that would resolve the issue. And that is my point: the argument has left the realm of conventional science since, except for restricted cases such as those dealt with by Hamilton, one must, as Barash admits, rely on intuition rather than experimentation. Thus a sociobiology that depends heavily on the maximization of inclusive fitness cannot convincingly lay claim to being scientific in the conventional sense.

Sociobiology and Sex

Darwinian theory has appeared again and again as ammunition in a social struggle over "the woman question." Should women go to college? Should they vote? Should they work outside the home? The storeroom of biological knowledge has undergone constant assault from raiding parties on both sides of the issue.[44] In addition to the "maximization of inclusive fitness," another weapon in the sociobiological arsenal is the concept of sexual selection. It all started with what Darwin viewed as the mystery of the universality of male excellence. Why should males be stronger and have larger antlers, horns, canine teeth, and on top of all that be the ones to sing beautiful courtship songs, to grow spectacular tail feathers, and to engage in strange behavior such as the mating displays of woodcocks and the intricate nest building of the bower and weaver birds? In some cases, such as that of the peacock, one might even expect natural selection to have eliminated the trait in question. Cannot a fox more easily spot and catch a peacock, brilliant as he is and weighed down by such an unmanageable tail, than he can a well-camouflaged peahen? How indeed did the peacock get to keep his feathers?

The answer to the problem, one emphasized strongly by modern sociobiologists, was something Darwin labeled *sexual selection.* Sexual selection has two components, intermale competition and female choice. Darwin recognized that large, specialized structures such as antlers in male deer were unlikely to have become common due to natural selection alone. Deer, for example, do not usually use antlers to protect themselves from attack, but depend more heavily on protective colors, alertness to surprise, and fleetness of foot—all characteristics likely to have become established through natural selection. What purpose then do antlers serve? Darwin saw but one—to enable males to spar with one another for mating rights with the females. In many species of deer there *are* ritual sparring matches each mating season. Darwin reasoned that the most successful combatants would get to mate earlier and more frequently, thus standing a better chance of passing on to the next generation those traits enabling them to win the contest in the first place. Too sophisticated a biologist to suppose that only antler size counted, he realized that sexual selection via male-male competition must have chosen a complex of traits which also included body size, vigor, and persistence.

Sexual selection was a complicated affair depending not only "on the ardor in love, courage and the rivalry of the males" but also on the "powers of perception, taste and will of the female."[45] For Darwin, female choice was key. Of what use would be a ritual "victory" if the female went off and mated with the loser? Postulating active female choice enabled one to understand that foxes have not eliminated long-tailed peacocks because the peahen's preference to mate with the most spectacular males has kept the genes for elaborate tail development present with high frequency in the population. Here is a case of balance between sexual selection and natural selection, predation working against males with spectacular tails, female choice working for them.

Darwin recorded many examples in which the male was more brightly colored or more highly specialized than the more juvenile-looking female. Extrapolating from these he suggested that in an evolutionary sense females represented the less variable, more juvenile or primitive state of the species. During evolution the male built more elaborate structures upon the simpler female baseline. (This idea is not unlike the embryological notion that the baseline of fetal development is female, discussed in chapter 3.) Not all nineteenth-century biologists bought Darwin's view. Alfred Russel Wallace, the most important nineteenth-century scientific interpreter of Darwinian theory, objected both to the idea that males were modified from a female base *and* to the notion that female choice played an important role in sexual selection.[46] Instead he felt that male structure and colors served as the starting point from which the different female coloration derived, mediated by natural rather than sexual selection. Although he believed in sexual selection involving male–male competition, he held that dull-colored females evolved because their drabness provided good protective coloration. Over the millennia, no doubt, brightly colored females raised fewer offspring because predators spotted them more easily. Nature thus selected for drably attired females, leaving the brilliant males to offer beauty to the world. Wallace's dismissal of female choice found easy acceptance, and the idea of a female role in sexual selection did not regain popularity among biologists until the early 1970s.

Wallace's and Darwin's descriptions of the animal world of male and female reflected widespread views about women. The Darwinian female represented the stabler but also the more juvenile form of the species. Despite her own passivity, she played an active role in the establishment of many a spectacular male modification. Interestingly, Wallace's investment in the notion of female passivity made it impossible for him to buy the concept of active female choice in sexual selection. And unable to make that purchase, he had then to stock up on the idea that the male represented the basic species type from which the less brightly colored female deviated.

The reopening of the Darwinian debate, engendered by modern attempts to explain animal altruism, contributed to an intellectual climate in which reassessment of Darwin's proposals about sexual selection became inevitable. Huxley and Darwin both clearly separated natural from sexual selection. Huxley cited the example of elephant seals, in which the males are a great deal larger than the females. Both sexes reach far greater size than other seal species; these interspecific differences, Huxley felt, must be explicable in terms of natural selection. Given the life-style of the elephant seal, increased size must have provided some selective advantage over smaller forms. But the striking size difference between the sexes could not have evolved solely by the action of natural selection on the species—male and female—as a whole. Male elephant seals fight intensely with one another for control over large groups of females. The winners get to mate with large numbers of females, the losers with none at all. For elephant seals (but not, for example, for other harem-forming mammals*) size is an important determinant of victory. The genes for increased size thus remain under heavy

---

* Jane Lancaster objects to the use of the word *harem*. She refers instead to one-male groups as those in which females are virtually self-sufficient, except for fertilization.[47] Note that this completely alters the significance attached to the "fact" of a particular social organization.

selective pressure and "the size . . . will tend to be pushed beyond the optimum, or what would be the optimum for other reasons."[48]

Sexual selectionists from Darwin to Edward O. Wilson have sought to explain the evolution of courtship displays. During such rituals males often strut about and show off any special courting colors (feathers, neck flaps, and so on) with great energy. Such behavior serves more than one function. To begin with it attracts the attention of members of the opposite sex. After the initial pairing, it helps to synchronize the couple's physiological readiness to mate: via the brain, visual stimuli can affect the pituitary gland to stimulate ovulation and thus increase the likelihood that fertilization will follow mating. It certainly wouldn't do, for example, for a male fish to shed his sperm into the open ocean *before* the female releases her eggs. Their courtship routine ensures that both eggs and sperm will end up in the right place at the right time. Natural selection was probably a major mechanism in the evolution of courtship displays, but some of the extreme displays such as the peacock's tail may well have become exaggerated through sexual selection—a selection that probably operated most strongly in species with strong intermale competition. Dr. Ernst Mayr, a highly respected modern Darwinist, sums it all up by explaining that the development of courtship displays and structures

> . . . was probably favored originally by natural selection to synchronize the physiological state of the two sexes, but . . . sexual selection is presumably super-imposed in all cases in which a male may gain reproductive advantage owing to an extreme development of an epigamic [courtship-related] character.[49]

Wilson describes the same events in different language. He uses the word *epigamic* to describe "any trait related to courtship and sex other than the essential organs and behavior of copulation":

> Pure epigamic display can be envisioned as a contest between salesmanship and sales resistance. The sex that courts, ordinarily the male, plans to invest less reproductive effort in the offspring. What it offers . . . is chiefly evidence that it is fully normal and physiologically fit. But this warranty consists of only a brief performance so that strong selective pressures exist for less fit individuals to present a false image. [The female] . . . will . . . find it strongly advantageous to distinguish the really fit from the pretended fit. Consequently there will be a strong tendency for the courted sex to develop coyness . . . its responses

will be hesitant and cautious in a way that evokes still more displays and makes correct discrimination easier.[50]

Two things strike me about this passage. First, the role of natural selection in the evolution of courtship displays, clearly emphasized in Mayr's description, is lost from view in Wilson's. Sexual selection suddenly becomes the total picture, the *only* description of the evolution of courtship behavior. Second, the metaphor of consumption leaps off the printed page. "Salesmanship," "sales resistance," "invest," "warranty": the words jar compared to Darwin's lyric that females selected the most beautiful males. In a century's time the image of the female consumer shopping for the best mate has replaced that of the protectress of beauty and refinement. Missing, too, is the image of two sexes cooperating to bring forth the next generation. Instead Wilson presents courtship as a contest in which males and females are fundamentally at odds; *his* interest is to reproduce regardless of how fit he is, even if success entails trickery, and *hers* is to increase her fitness by exposing his deceit, choosing for herself the best possible mate.

Sociobiologists offer a simple physiological reason to explain why males and females have such different approaches to "maximizing their inclusive fitness." Females produce eggs, large cells stuffed full of nutrients which support the fragile life of the developing embryo. In an apparent trade-off between size and quantity, the females of most species produce relatively few such giant cells. A single female mammal, for instance, may make as few as four hundred in her lifetime. Males on the other hand produce millions of sperm, little bitty cells that stimulate egg development and transfer a set of the father's chromosomes to the new generation. Citing as evidence the difference in size and number of male and female sex cells, sociobiologists claim that females must invest more physiological energy in putting together an egg than males do in making a sperm. As an aside to the main explanation of their argument, I must point out that they provide no evidence to support this idea, apparently believing it to be obvious to even the most casual observer. How one weighs the energy costs of judicious manufacture of a small number of rather large, well-stocked cells against the profligate production of millions and millions of tiny ones, though, is far from clear. Be that as it may, the sociobiologists lean heavily on this nonaxiomatic axiom. A female, they say, starts out investing more in her sex cell and thus must protect her startup costs by making sure the egg gets the best sperm possible and by guarding the fertilized egg. If one adds on to the situation internal development and prolonged postnatal care, the differential becomes enormous. Here's how David Barash puts it:

> [The female] has much more invested in each of those eggs than [the male] has in his sperm and this asymmetry is particularly apparent in higher vertebrates. . . . male birds and mammals produce sperm in incredible abundance with each ejaculation, and these sperm can be readily replaced. Males can usually walk . . . away from the consequences of a copulation. . . . The consequences of a bad decision . . . fall particularly heavily upon females and hence it is not surprising that they tend to be the more discriminating sex.[51]

At the time of fertilization females have already put more energy into gamete production and thus, so the theory goes, are less likely to abandon ship. In addition, females of many species carry the embryo inside them, or guard eggs while they develop. With passing time the energy invested in producing and then rearing to birth a fertilized egg increases the difference in female versus male investment. Thus, according to sociobiological theory, a male can benefit by abandoning females early in the game, when he has invested little and she a great deal.

On the physiology of gamete production Barash and others have built entire edifices of behavior. This particular aspect of sociobiological theory bears intriguing similarities to one from the turn of the nineteenth century expounded by Patrick Geddes, a zoologist turned sociologist, and his student J. Arthur Thompson.[52] They argued that the sperm exhibits a catabolic or active quality while the egg is anabolic, passive, and well fed. From these differences in the metabolism of the male and female sex cells they deduced a host of behavioral differences found in adults: males are smarter, more independent, and more energetic than their patient, intuitive female counterparts. Although the sociobiological theory uses more up-to-date biological terminology and emphasizes courting and child-rearing behaviors rather than more general qualities such as intelligence, its biological content is no less naive than Geddes's and Thompson's eighty-five-year-old conceptualization.

Modern sociobiological theorists start with the gametes and end up explaining why females tend toward monogamy and males toward polygamy. Generally, they say, a female's reproductive physiology limits the number of offspring she can produce; she increases her fitness by ensuring that her young stand the best possible chance of survival and reproduction. Since in theory a single male can produce many more offspring than a single female, the quality of any individual offspring may not matter so long as he gets to make a whole lot of them. Thus males will mate without much discrimination and as often as possible, and any female will do, so long as she's young and healthy. Once a female has mated, though, she drops out of the running because her investment in the egg dictates that she take care of it rather than running off to mate again. So it pays for her to be highly selective, putting off the male until she can be sure he's a good choice. That polygyny is more common than polyandry, say sociobiologists, offers support for their assertions.

That polyandry exists at all, though, presents a puzzle. Certain shore birds, for instance, completely reverse sex roles. The female is larger and more brightly feathered and is the aggressive suitor. After mating with one male she leaves him behind to sit on the eggs and feed the hatchlings, while she moves on to a second and sometimes even a third round of mating and egg laying. The males in each case perform the child-care function.[53]* Sociobiologists rationalize the existence of polyandry (which has evolved at least five different times in birds) by reference to another special invention, an account of sexual selection in terms of something they call "parental investment." "The relative parental investment of the sexes in their young," writes Professor Robert Trivers (who invented the idea), "is the key variable controlling the operation of sexual selection." Trivers hoped to offer a general theory to explain the wide variety of mating and child-care arrangements found in the natural world. He suggested that any time one sex (usually the female) invests more in reproduction and parenting, members of the other sex will compete among themselves (sexual selection) for mating privileges. The patterns seen in the world today, he suggests, were historically influenced by the differences in initial energy investments in differently sized sex cells. He defines parental investment as "any investment by the parent in an individual offspring that increases the offspring's chance of surviving (and hence repro-

---

* It is interesting to note that the females from these species produce more testosterone in their ovaries than do the males in their testes.

ductive success) at the cost of the parent's ability to invest in other offspring."[54] Investment starts with the production of the sex cell and ends when the parent is shed of its young.

Trivers plays out the consequences of this theory in a number of ways. Since sexual selection occurs in cases for which the total parental investment of each sex differs, these cases ought always to result in greater sexual dimorphism, that is, one sex being larger or more brightly colored than the other. Similarly, in species that are not particularly dimorphic, the amount of parental care should be roughly the same for both sexes. He tests his hypothesis that the relative amount of parental investment controls sexual selection with the example of polyandrous birds, saying, in effect, that they are the exceptions that prove the rule. For in this rare example in which the female invests less in parental care, she is the one to show the gaudy effects of sexual selection. Although this conveniently fits into part of his scheme, it does not explain how polyandry ever evolved to begin with. Female birds from polyandrous species, after all, do produce eggs of enormous size and the males make tiny sperm. Using the gametic energy investment theory, polyandry ought not to happen. And surprisingly, neither Trivers nor Wilson—who offers an extended account of Trivers's ideas— seem worried about it.

The polyandry perplexity not withstanding, Trivers has a lot to say about the relations between the sexes. He develops his ideas about parental investment to explain everything from philandering males who desert their mates and the sexual double standard to the deep humiliation felt by a cuckolded husband. True monogamy, he writes, can't really exist because "male sex cells remain tiny compared to female sex cells. . . . The male's initial parental investment, that is his investment at the moment of fertilization, is much smaller than the female's, even if later, through parental care, he invests as much or more."[55] Thus a male, even a monogamous one, remains ambiguously committed because at any moment he can abandon an impregnated female for a new mate and still come out ahead in the energy investment contest. The female is stuck with three choices. She can abandon her already considerable investment (ditch the kid) to search for a more faithful male. She can raise the young alone; if she succeeds, the male who abandoned her lucks out—genetically speaking—and he may have managed to reproduce once or twice again in the meantime. Finally, the female can try to deceive another male into helping her out, a possibility with some likelihood of success if she doesn't already have eggs or offspring to show for her troubles.

Trivers also offers us new insights into cuckoldry, a word derived from the habit of the female cuckoo of laying her eggs in the nests of other birds. The cuckoo hatchling invariably heaves its noncuckoo nestmates over the side of the nest and becomes the sole, enormous, ugly child of its still-doting foster parents. In modern usage a cuckold is the husband of an unfaithful wife—a far nastier and more humiliating state, apparently, than being the wife of a philanderer, for which in fact no word exists. In any organism in which fertilization is internal the male cannot be sure whose sperm did the fertilizing, so "to the degree that he invests in the care of the offspring it is genetically advantageous for him to make sure that he has exclusive access to the female's unfertilized eggs."[56] Thus selection would favor the evolution of males that guard their females through territorial exclusiveness, dominance, or the development of long-term monogamous bonds. A cuckolded husband is far worse off, genetically speaking, than the wife of an unfaithful husband. She at least is certain that the children she bears are her own; he must take her word for it. In Trivers's view this biological fact makes plausible the double standard and helps explain the severity of sanctions taken against sexually adventurous women.

Parental Investment: What Is the Evidence?

Trivers bases much of his work on animal studies. Yet neither he nor other sociobiologists are timid about applying their ideas to humans. Wilson writes:

> The principal significance of Trivers' analysis lies in the demonstration that many details of courtship can also be interpreted with reference to the several possibilities of maltreatment at the hands of the mate. The assessment made by an individual is based on rules and strategies designed by natural selection. . . . potentially the implications for the study of human behavior are very great.[57]

At one level it all seems quite plausible. Using the thread of internal consistency, sociobiological analyses bind together many seemingly disparate facts. A good explanation ought certainly to do just that—but, alas, *just* that is not enough. Viable theories require experimental testing in order to remove them from the realm of fantasy and place them in the world of science. Since evolution by natural selection requires favoring some genetic variants over others, the first and foremost criterion is the demonstration that the trait of interest is really under genetic control.

Attempts to demonstrate genetic control, however, encounter a number of serious difficulties. Consider two very different species of fish. One, the codfish, produces millions of eggs, spilling them without further attention directly into the open ocean, where the vast majority die. Another, the eelpout, produces only a few offspring which the female bears alive. Using the concept of parental selection one could explain the difference between these two fish, suggesting that the codfish adapted by producing large numbers of eggs while abdicating all parental responsibility. The eelpout, in contrast, invested more heavily in each egg but "saved" energy by producing fewer of them. The total energy cost for each species might be the same, the two life-styles taking the appearance of coequal adaptive choices.

It might have happened that way, but it needn't have. In fact it would *only* have happened that way if the ancestral populations from which codfish and eelpouts evolved had equally variable gene pools for the traits of egg production and maternal care. If, however, the ancestral codfish population had a fair amount of genetic variation in egg production but relatively little variation with regard to maternal care, then it could not have chosen between two equally possible reproductive strategies. Instead natural selection could have acted only on egg production, since this was the only trait with genetic variants from which to choose. As the critic from whom I have taken this example put it, "Knowledge of the relative amounts of genetic variance for different traits is essential if evolutionary arguments are to be correct rather than simply plausible."[58] The example of the codfish and the eelpout, by the way, resembles that of the one- and two-horned rhinoceri, mentioned earlier in the chapter.

Another problem is that one cannot use knowledge about the genetic variance of a particular trait in a currently existing species to make accurate inferences about the trait's evolutionary origins. Consider further the evolution of the ancestral codfish. If there were a great deal of genetic variation for egg production and a strong likelihood that cod eggs would be snapped up by waiting predators, then those females genetically disposed to lay lots of eggs might have few surviving offspring, but they would in turn pass along the genes for greater egg production. At the same time, the genetic variants that produced few eggs would die out and that genetic variation would be lost from the population. Although it could be reintroduced by mutation or migration from a neighboring community, the net effect of selection would be to reduce the genetic variation in egg production. In fact a fundamental rule of natural selection is that the more highly selected a population, the lower the amount of genetic variation it contains. This accounts for the fact that animals that become too perfectly adapted face the probability of extinction, since as a population they lack the genetic resources to respond to changes in the environment.

Suppose that a sociobiologist wished to test his or her hypothesis that the codfish reproductive mode represents one end of a spectrum of possible adaptive choices open to fish. He or she could measure genetic variability in egg production in present-day populations. Finding little, the investigator could argue for the implication that in ancestral populations there must have been a great deal of variability, all eliminated in the interim by natural selection. On the other hand a finding that populations of codfish contain genetic variation for egg production would show that different levels of egg production are inherited and thus potentially susceptible to natural selection. Both results can be used to favor the idea of strategic choice in natural selection. In fact, "there is no conceivable observation about genetic variance at present that could prove the contention of past evolution of the trait."[59]

So how *can* one prove things about evolutionary history? In the case of the codfish one might, if exceedingly lucky, come upon some fossils in which soft parts were preserved. Such preservation happens rarely—most fossils consisting of mineral replacements of hard structures such as bones. But granting the possibility for the moment, it might be possible to find fossils of protocodfish which carried very different numbers of eggs and to gain from them some sense of how variable the ancestral population might have been. There is, unfortunately, no direct fossil record of behavior, no way to know whether the protocodfish exhibited variability in the amount of parental care lavished on its young. "Fossil behaviors" can only be inferred from physical evidence in the fossil record, and such interpretations must necessarily include considerable guesswork. Thus the sort of theorizing done by Trivers and others contains within it fundamental problems of proof. Stories about past evolutionary events will be the hardest to support, while assertions about behaviors seen in the wild today will at least be open to experimental examination. One such behavior, to which we will now return, is the animal matings which some biologists have designated as rape.

Rape: Reprise

Thornhill hypothesizes that some male flies force themselves on females only when they have lost out in competition for nuptial food offerings. Since possession of a food source provides incontrovertible physical evidence—the male cannot easily deceive a female about his fitness. Failing in competition for food leaves male scorpionflies with two choices, rape or consignment to extinction. Such an analysis implies that behaviors which enhance reproductive fitness will be favored by natural selection. That rape in scorpionflies has evolved for just this reason becomes the unstated conclusion.

Although it is possible to test the *plausibility* of Thornhill's conclusions, he has not done so. His follow-up studies pursue quite a different tack. A first step would be to make sure forced copulation is a natural rather than an experimenter-induced event, for if it is but a laboratory artifact it loses all interest. On this point Thornhill's own writing is unclear. In one paper he reports observations of rape in nature in seven species of scorpionfly and in "most of the 18 species" he has studied in the laboratory.[60] But in another he mentions low frequencies of forced copulations in field enclosure experiments, while in open field experiments he observed no successfully completed forced copulation.[61] How robust (to use a favorite word of scientists), one wonders, is the claim that scorpionflies rape in the wild?

Leaving aside the actual biological importance of the observation, we can consider alternate explanations that are not based on fitness strategies. Thornhill mentions that males secrete a substance which attracts females to their food cache. Do females secrete a substance which stimulates male mating behavior? If this were the case it might be that most males would try to mate with any nearby female, although females resist matings with males that have no food to offer. Indiscriminate attempts to mate might occasionally succeed but would not represent a special adaptive strategy on the part of foodless males.

For rape to have evolved as an adaptive strategy it must be heritable. And if it were heritable it would be possible to raise up the offspring from females which have experienced forced copulation and to test their sons for such behavior. Do these sons also rape while those of "nonrapist" males do not? As the discussion of codfish and eelpout evolution points out, such a study would not prove that Thornhill was right about the *origin* of the field behaviors he first observed; but such experiments could at least make his theory minimally tenable.

If the scorpionfly studies leave something to be desired, Barash's work on mallards is far worse. He expands on the question of rape in ducks by worrying about the possible options of the victim's mate. Following observations of forced copulation in a population of ducks living in "a semi-natural urban environment," he recorded and interpreted the responses of the females' mates using "the sociobiologic theorum that animals should behave in ways consistent with maximizing their inclusive fitness." He suggests that a drake has two choices when "his" hen is attacked. Either defend her and drive off the attacker, or try to introduce his own sperm right after in the "hope" that the sperm would compete with those of the assailant. Barash's unstated assumption is that both the attacking behavior and the mate response represent evolved adaptation, and must therefore have a genetic basis. Since he observed both of his predicted behaviors, he felt justified in reaching the following conclusions:

> An evolutionary perspective on behavior suggests that individuals will behave so as to maximize the difference between the benefits and costs associated with any potential act, with both benefits and costs evaluated in units of inclusive fitness. Rape of one's mate imposes a potential cost in that it increases the likelihood of another individual's fathering her offspring. The responses available to a rape victim's mate also carry benefits and costs and the observed pattern suggests that the mate behaves in accord with evolutionary prediction.[62]

Barash bases his discussion solely on what he finds to be consistent with the evolutionary possibilities he imagines. If the scheme fits, wear it; the evidence (or lack of it) be damned. Most unsettling, though, is his disregard of information pointing to a very different analysis of forced copulation in mallards. In fact, following the publication in the journal *Science* of his mallard article, two separate letters to the editor appeared, criticizing both the inadequacy of his data and, more fundamentally, the usefulness of the very phenomenon of "rape" for the sort of evolutionary analysis he attempts.[63]

As it turns out, rape in mallards is pathological. One careful

191

study of ducks living under abnormal, high density conditions did indeed show some of the same behaviors observed by Barash. But they were far from adaptive. In fact the overall fertility in this particular crowded population was very low and the reproductive significance of violent mating attempts uncertain.[64] Barash, who provided no data on the density or the actual breeding success of the population he studied, dismisses the criticism that rape is a pathological response to overcrowding by saying that he certainly "would welcome studies in relatively undisturbed habitats." Barash cites one other paper as evidence that rape is common in ducks. This study too was done on urban ducks, but under relatively healthy conditions, and the cited researcher does indeed observe "rape" attempts. But Barash fails to report a second key observation. The attempts were seasonal and only occurred at the very *end* of the breeding season. Thus they could not have had anything to do with increasing reproductive fitness, and seemed instead to be associated with the seasonal breakup of the breeding pair.[65]

In sum there are two quite extensive studies reporting forced copulation in mallards. One comes from a highly overcrowded population exhibiting a wide variety of pathological behaviors and low fertility (a fact which Barash fails to point out in his citation). It is well known that stress in animal populations can bring about aggressive and destructive behaviors, none of which result from natural selection for some alternative reproductive strategy. The second study suggests that forced intercourse may be an aggressive behavior associated with the yearly breakup of the breeding pair. Since it only shows up after reproduction is over for the year, it too fails to correlate with increased reproductive fitness. In other words, a quick check of Barash's "supporting" references makes clear that forced intercourse in mallards has nothing whatsoever to do with alternate reproductive strategies. The data render Barash's neat little scheme bankrupt.

Although Thornhill originally presented his ideas on human rape in underdeveloped form, he and Nancy Thornhill and William and Lea Shields recently published extensive accounts offering "an evolutionary perspective" in the same issue of a scientific journal. They uncritically cite the work on animal rape and go on to draw conclusions about humans. (The Thornhills' article was discussed earlier in this chapter.) The Shields suggest that human males employ one of three tactics to reproduce. The first of these, cooperative bonding, appears in our marriage system; the second, manipulative courtship, involves deceitful seduction, i.e., mating and running. Finally, if these tactics fail, forcible rape remains an option. Here is how they state their case:

> Illustrating that rape is likely to be adaptive for males does not imply that *all* males will rape *exclusively* or *all* of the time. Intuitively, we feel that *individual humans* that possess *all three* of these mating tactics, each used in appropriate circumstances so as to minimize potential costs in a conditioned strategy, are likely to be more successful than males with a pure strategy (i.e., all males either rape *or* court honestly) and more stable than any polymorphic population (some males rape exclusively, others only court honestly).

Plainly put, all men have the potential to rape, and may be expected to do so if they can get away with it:

> We suggest that *all* men are potential rapists. . . . We expect that the probability of a particular individual raping will be a function of the average genetic cost/benefit ratio associated with the particular conditions he faces.[66]

These authors do not argue that men consciously think through the possible genetic advantages of rape. Rather, they view this non-conscious calculation as an "ultimate cause," one that originated early in our evolutionary history. Rape, they acknowledge quite freely, may have "proximate causes" such as aggression, the desire to humiliate, or sexual gratification—just as many feminists have argued. But those proximate causes, say the Shields, represent evolution's way of carrying out its ultimate desire of maximizing genetic fitness.* Again, quoting from their article:

> Ultimately men may rape because it increases their biological fitness and thus rape may serve, at least in part, a reproductive function, but in an immediate proximate sense it is as likely that they rape because they are angry or hostile, as the feminists suggest.[67]

The use of the ultimate-proximate distinction is clever, because it is totally unassailable. *Any* motivation for rape, regardless of how little it has to do with reproduction, can still be explained on an

---

* They also explicitly exclude nonadaptive rape, especially rape in which the victim dies, homosexual rape, and presumably—although they neglect it altogether—gang rape. They argue the legitimacy of the exclusion by saying that these forms of rape are rather rare compared to your run-of-the-mill daily rape.

evolutionary basis by arguing that it is the proximate effector of the ultimate genetic cause!

All of this would be merely annoying were it not for the fact that it has immediate practical application. The Thornhills and the Shields deplore rape, but feel that society can figure out how to overcome its evolutionary heritage and put a stop to this morally heinous activity only if it acknowledges the accuracy of their theories:

> Our model . . . implies that reducing sexism, "raising the economic and social status of women and the consciousness of men," and rehabilitating the rapist . . . may have little more than minor effects on the incidence of rape. Only when efforts are directed at altering all of the genetic and environmental factors important in the natural control of rape would we predict success.[68]

Just what efforts do they suggest? If we set the cost of rape high enough, then the practice ought to diminish. One possibility is to increase drastically the severity of punishment for rape. This, the Thornhills acknowledge, might be difficult, especially given the trend away from life sentences for rape. As the Shields put it: "The application of the severest forms of retribution (e.g., execution or castration) . . . might carry many ethical, practical, and political problems."[69] The next best thing, then, would be to ensure *some* punishment; to stop rape, they suggest, we must achieve a high rate of criminal conviction complete with jail terms. This and this alone is what would work, according to the evolutionary hypothesis; yet how to achieve higher conviction rates *without* changing the social and economic status of women never seems to become an issue for these "theorists."

Neither the Thornhills nor the Shields discuss an alarming corollary to their evolutionary hypotheses, one that already plays itself out within the criminal courts. In fact they would probably refuse to acknowledge the relationship between their ideas and the releasing of rapists without jail terms on the condition that the convicts take the drug depo-provera to reduce their sexual drive. Yet this has happened. Item: A forty-one-year-old man convicted of raping his stepdaughter over a period of seven years received a one-year jail sentence under the condition that he take a drug to reduce his sex drive.[70] (That man, the wealthy heir to the Upjohn fortune, apparently represents a counterexample to the Thornhills's view that financially successful men don't rape.*) Another rapist, convicted of raping the same woman on two different occasions and attempting it on a third, escaped jail altogether after his lawyer successfully argued that he suffered from uncontrollable biological urges. Instead the jury sentenced him to take depo-provera to control his need to rape. His lawyer argued that jail was not appropriate at all because the victim "came across consistent with what she underwent, which was not a particularly brutal rape; as rapes go, a relatively mild rape." A physician who touts the use of depo-provera to control rape says that some men have "unbearable physical urges. They have all been found to have unusually high levels of testosterone in their blood."[72]

If I were a clever (to say nothing of unscrupulous) lawyer, I would be heartened by reading the Shields's and Thornhills' work. Within it lies the ultimate evolutionary explanation for the supposed existence of men with strong biological urges to rape and the scientific basis for defending them when they do. If the builders of unfounded evolutionary theories about rape do not foresee how their work will be used and if they claim that it is not their fault, that they simply injected their hypotheses into the free marketplace of ideas where they could be tested and rejected if wrong, then they are at best fooling themselves. At worst, they engage in the most irresponsible sort of academic navel-watching. Rape, however, is not the only human activity about which sociobiologists write. Their ideas about human sexuality are far-reaching.

The Human Biogram

Human sociobiology is a theory of essences. Wilson sets as his task the tracing of human qualities back through evolutionary time. "The exercise," he writes, "will help to identify the behaviors and rules by which individual human beings increase their Darwinian fitness through the manipulation of society. . . . We are searching for the human biogram."† In contrast to the uncertainty surrounding the chicken and her egg, it is clear that the human biogram precedes

* Public outcry in this case forced the judge to reconsider his initial ruling. It is perhaps ironic that Upjohn is the world's largest manufacturer of depo-provera, which—despite the fact that it is banned in the United States because it induces cancer in test animals—Upjohn sells to third-world countries as an injectable birth control drug.[71]

† Biogram is a term used by Wilson; biological program is its basic meaning.

all other arrangements. "Our civilization is jerrybuilt around the biogram," asserts Wilson.[73] And by stripping away the cultural trappings characteristic of human societies, we can lay bare the human essence, that underlying genetic something which, after all is said and done, profoundly influences the social structures developed by members of the human race.

On the essence of male and female, sociobiologists have no dearth of opinions. To start with, men—more or less universally—are more sexually active than women, making it no surprise that among humans polygyny is far more frequent than polyandry. Then, too, men are less responsible, less nurturant, and less emotionally expressive,[74] while women are coy and sexually selective. All humans live in families with certain universal traits: in "an American industrial city, no less than a band of hunter-gatherers . . . during the day the women and children remain in the residential area while the men forage for game or its symbolic equivalent in the form of barter or money."[75] And, they tell us, while both males and females experience sexual jealousy when faced with an adulterous spouse, it "may be a more or less 'innate' response among husbands and a learned response among wives." Furthermore, the male desire for sexual variety is "natural and universal."[76] The male is more aroused by visual stimuli than is the female. Males are more naturally attracted to young and therefore highly fertile women. Since a male's fertility remains more or less constant over long periods of his life, youth is not a primary criterion for attracting the female. Instead she exhibits something Wilson labels *hypergamy*, the "female practice of marrying men of equal or greater wealth and status."[77] Such behavior follows from the existence of male–male competition. In nature, after all, most females get to mate, but whenever there is intermale competition the weaker, less fit males lose out. Wilson sees a human counterpart in societies that practice female infanticide among the upper classes. Sons survive and marry lower-class (hypergamic) females who—along with their dowries—move continually upward. The breeding system almost completely excludes poor males.[78]

When coming up with "universal" traits requiring explanation, many sociobiological accounts of human interactions so ignore the complexities of behavior and culture and make such selective readings of the literature, that it becomes a challenge to respond in any fashion other than unabashed astonishment. Although the problem of correctly describing human behavior is a serious one, the deeper issue lies with the very notion of human essence—not with the idea that there is a human nature, but with the thought that there is a *particular* human nature, visible when all culture and learning is stripped away.

Desmond Morris, in his book *The Naked Ape,* described human behavior as deriving directly from our ape-like biology. Most sociobiological writers would disagree—or so they say. Here's how Trivers puts it: "Sociobiology emphasizes facultative traits, based on genes, which permit individuals to adjust themselves . . . to a great array . . . of contingencies. . . . The fact that we are highly evolved social actors means that each of us has the genes with which to play almost all of the roles."[79] Wilson suggests that only 20 percent of what we are is attributable to genetic bias, 80 percent to cultural overlay. Trivers and Wilson certainly sound more reasonable than Desmond Morris, but what do they mean? If sociobiologists have discovered that humans can behave as both loving and rejecting parents and can be both violent and peaceful and can cheat and lie as well as cooperate, then they have said nothing that is not obvious to the most casual observer. The Quaker and the sniper coexist in every city in America. But sociobiologists say a good deal more than that: they say that what is universal is natural. And their universals include the double standard, sexually aggressive men, the sexual division of labor, conflict between male and female and between parent and child, and families in which men bring home the bacon while women care for the kids. We must consider the claim of essence from three interrelated viewpoints: (1) Is there any supporting evidence for the idea of universal human behaviors? (2) Do supposedly universal behaviors have the same meanings in different social contexts? (3) If there is some meaningful human universal, does that then imply its evolution through genetically based adaptation and natural selection?

In the radio and television series "Dragnet," Jack Webb kept asking for the facts ("the facts, Ma'am, nothing but the facts"). And perhaps nowhere are the facts murkier than in the study of human universals. Even establishing that something as simple as the human smile is inborn rather than learned, and that it signifies happiness or pleasure in all human societies, is a complex undertaking. That it is probably unlearned can be ascertained from studying deaf, dumb, and blind children, who will smile quite unmistakably. That the smile is a sign of happiness can be ascertained by a variety of cross-cultural studies. But it is no easy task.[80] Among the behaviors considered by sociobiologists, something resembling agreement among groups as diverse as sociobiologists, feminist anthropologists, and Marxist sociologists exists about one: the division of labor by sex. With the exception of small groups such as the newly discovered Tasaday, all human societies seem to have some form of division of labor by sex. The problem is that no two societies have exactly the *same* form.[81] What's good for the gander in one culture is just fine for the goose in another. Nor is it true as Wilson claims that men always earn the bread while women raise the children. In some cultures women are the merchants and financiers; in others men do the bulk of the child care. In fact how to interpret the sexual division of labor is a matter of considerable debate[82] and, although they never acknowledge their part in it, sociobiologists simply hold up one end of a diverse spectrum of thought on the topic.

The division of labor by sex embodies a seeming contradiction: it is a human universal but it has no universal meaning. Instead each culture has its own particular division of labor by sex (one that may be more or less rigid) and attaches to it its own set of interpretations. There may well be some human essence that leads to a division of labor by sex, yet there seems to be nothing in our nature which says what that division will be or how we will translate it into cultural meaning. It seems then that we can extract meaning only by examining the division in a particular social setting. But if this is true, what becomes of Wilson's biogram? I argue that it turns into one of two things—a tautology or (despite protests to the contrary) a prescription for human behavior.

Any harmony that might result from agreement that most cultures have a sexual division of labor dissipates completely when one looks at any of the other sociobiological "universals." Barash, for example, writes that sexual attractiveness to one's mate is a universal. One need only take note of cultures in which marriages are arranged at birth to question the accuracy of this generalization. But even if we were to grant the assertion we would still have to question its significance. There is certainly no generalizable standard of female beauty. In some cultures men find heavy women more sexually appealing than thin ones. In others the reverse is true. Within the Western European cultural tradition, the standard has differed in different historical periods. Barash would argue, though, that the specific standard is unimportant—that there *is* a standard is what counts. But one is then at a loss to know what he means. This fallback stance seems rather to return the argument to a tautological status. It is a little like holding up a pail before a thirsty person. "It is full," one might say. "With what?," would come the reply. For a person in need of a drink a pail filled with diesel fuel holds nothing of interest.

By insisting on understanding the meaning claimed for a human universal, it fast becomes clear that there is no single, undisputed claim about universal human behavior (sexual or otherwise). The notion of a naked human essence is meaningless because human behavior acquires significance only in a particular social context. This is the case even for clearly "biological" behaviors such as eating and drinking. All humans do both to keep alive. The ability to experience thirst is localized in a particular part of the brain. Yet the act of drinking has many meanings. We go out to a bar with friends not to quench our thirst but to be sociable. We raise our glasses in unison at a party to celebrate an important event, not to meet our physiological requirement for water. Even this act which *can* fill a biological need cannot be understood outside of its social context.

But suppose this argument is wrong, that one could identify a set of meaningful universal truths about human behavior. Then universality might indeed imply a human biogram, a muted genetic deep structure underlying the many-layered cultural overlay. And such a biogram might represent a set of genetically based adaptations to the selective forces that existed from one hundred thousand to four million years ago, when humans evolved from their primate cousins. But a human biogram might also have resulted from random genetic events which had nothing to do with adaptations and fitness maximization. Far too little is known about the origin of Homo sapiens to rule out such a possibility. Thus even if one were to grant a starting premise of human essence, it remains impossible to figure out which essences are adaptations arising under the pressure of natural selection, helping to increase fitness, and which just happened along for the ride. *Human sociobiology is a theory that inherently defies proof.*

Both of the above possibilities suggest a genetic basis for specific human behaviors. Yet a third option remains. Quite early in human prehistory protohumans could have learned certain behaviors and taught them to their offspring. If the present-day worldwide population of humans all evolved from a small progenitor stock (and on this point there is some, albeit far from unanimous, agreement), then certain kinds of behavior might be universal, yet learned rather than genetically programmed. Such an idea is not far-fetched. Today's primates learn social traits and teach them to the next generation. Rhesus monkeys raised in social isolation do not innately "know" how to interact with other rhesus. Nor do orangutan infants taken from their mothers innately know how to forage in the rain forest. They must be taught to do so by human surrogates before they can be returned to the wild.

Monkeys even have culture. They invent things and pass on their discoveries by nongenetic means. For example, a group of scientists spread grain on a sandy beach in order to attract wild Japanese macaques to a central location for observation. At first the members of the monkey troop tediously picked out the food, grain by grain. But one day a young female had a bright idea. She picked up a mixture of sand and wheat, rushed to the seaside, and threw it in the water. The sand sank and the wheat floated, and she simply scooped it from the surface. Her siblings learned the trick from her, followed by her mother and other juveniles. Males and older females without young were the slowest to pick up the new skill, but it is now handed down to succeeding generations. Yet an outside observer, seeing the universality of the event but not knowing its history, might hypothesize a genetic basis. (The inventor, by the way, was something of a monkey genius. Some years earlier she had learned and then taught the others to wash the sand from sweet potatoes left on the beach by the human investigators.)

Perhaps it doesn't really matter if a trait is genetic or learned. Maybe all that's important is that stable, widespread behaviors regardless of their cause have helped humans to survive and evolve. The whole debate may be much ado about nothing, but I think otherwise.

> Genes, being part of our bodies, live and die with us. They can be spread only through our offspring. Cultural templates are separate from our bodies. They can be multiplied, spread . . . during the lives of a single individual. Cultural evolution is therefore, far more flexible than biological evolution.[83]

Change, time, flexibility: in the controversy about genetic versus cultural evolution, these are the issues. Curiously the sociobiologists state with conviction that they are at one with their critics. Their cry is that in order to eradicate human social ills, we must first understand them. "Sociobiology," writes Barash, "helps to identify some of the possible roots of our injustice—male dominance, racism.... If any change is to occur ... we would do well to understand the biological nature of our species."[84] Trivers, too, bridles at the thought that what he calls "Darwinian social theory" is politically reactionary. To the contrary, he suggests that "the concepts of parental investment and female choice provide an objective and unbiased basis for viewing sex differences, a considerable advance over popular efforts to root women's powers and rights in the functionless swamp of biological identity."[85] Self-knowledge must precede change—a most plausible theme.

Sociobiologists writing about humans do indeed take up the challenge of social change. "The major interest most of us have in rape," writes Donald Symons, "is in its total elimination." This he suggests could happen in one of two ways. Males and females could be kept separate, as they are in certain Moslem societies, building in effect a structural barrier to prevent rape; or, young boys could be bred with social inhibitions that would "produce men who want only the kinds of sexual interactions that women want." The problem is that the former course of action is unlikely in our culture, while the latter entails a loss of freedom—a price that may not be worth paying. Without explanation Symons asserts that the rearing conditions needed to eliminate rape "might well entail a cure worse than the disease."[86] Worse for whom, one might wonder.

In his discussions of social change Wilson uses a similar strategy. After stating his sincere interest in change he discusses the pros and cons of several options for achieving it. Finally, after showing the pitfalls of each he reluctantly concludes that things really aren't all that bad as they stand. For example, he believes that slightly different inborn emotional dispositions account for the development of some of the "sexual universals" in which sociobiologists believe. Calling upon the hope offered by scientific truth, Wilson suggests that full recognition of these innate dispositional differences will help to light the path chosen by future societies: "In full recognition of the struggle for women's rights that is now spreading throughout the world, each society must make one or the other of the three following choices." Right now most societies "condition [their] members so as to exaggerate sexual differences in behavior." This, Wilson is well aware, usually leads to greater male domination and sexual inequality, although he holds out the hope that we could learn to safeguard human rights while savoring the richness of human difference. Alternatively, a society could bring up its children with an eye toward eliminating all sexual differences in behavior.

Wilson fully believes this would be possible and that it might result in a more just, productive, and harmonious society. But—and this sounds the death knell for behavioral identity—"the amount of regulation required would certainly place some personal freedoms in jeopardy and at least a few individuals would not be allowed to reach their full potential." Finally, he proposes that we simply open all the doors but provide no special help for women to pass through them.[87] But here, too, there are pitfalls, for even with identical education and equal access Wilson thinks that their biological biases will leave men more likely to dominate in political life, business, and science and women more likely to do the child care.

One is thus left in a state of bewilderment. None of the options for change is perfect and some seem downright appalling. How *do* we recognize the spreading struggle for women's rights? In Wilson's view the best bet is to live with our differences: "I am suggesting that the contradictions are rooted in the surviving relics of our prior genetic history, and that one of the most inconvenient and senseless, *but nevertheless unavoidable*, of these residues is the modest predisposition toward sex role differences"[88] (emphasis added). From change to the status quo, from options to determinism, in one easy lesson.

There are variations on Wilson's theme. In the last chapter of his book *The Ape Within Us* John MacKinnon muses about the "Apes of the Future." He concedes that the ability to limit family size makes it "surely right" that women "should be granted" a chance to work outside the home—equal opportunity and all. Still, the future Ms. Average will be less aggressive, more emotional, and "her biological make-up has . . . designed her for fulfilling quieter, less spectacular roles. . . . For all her opportunity and capability Ms. Average is going to end up in a supportive domestic role." With one difference: in the past she received special recognition for that role but in her liberated future such special attentions will be lost, and she may be "haunted by the misery that she ought to be making more of herself." That, it seems, is what social change brings. Women's liberation won't wash. "Women's libbers"—as MacKinnon calls them—"will be constantly let down by their sex's biologically lower motivation for fighting for glory in the industrial head-hunt of the economic rat-race."[89]

The solution to the dilemma our future Ms. Average will face seems to lie in revaluing male and female roles. Just how we should do this MacKinnon doesn't say. But if only women were to receive

202

more rewards for the wonderful things they do so specially, men and women could proceed with their various (complementary) activities. He proposes in essence a return to the two spheres of the Victorian era.

Sociobiological theory has already been put to social and political uses, for example in the movie *Doing What Comes Naturally*. With a background of voice-overs from interviews with Wilson, Trivers, and Dr. Irven de Vore (another outspoken sociobiologist), the movie shows male sheep locking horns, teenage girls and boys engaging respectively in coy and fickle behaviors, and scenes of massive bombings and battles in Vietnam. The message is so crudely put that the sociobiologists seen in the film publicly dissociated themselves from it.[90] But too late, alas. The film is still available, selling its message about female coyness, male aggression, and the inevitability of war. Others, too, have jumped onto the bandwagon. A psychiatrist has suggested that psychological disorders be reclassified using the concepts of adaptive strategies, survival, and inclusive fitness. "For example, a taxonomy might be developed which rated disorders in terms of their potential effects on inclusive fitness. A severely 'psychotic' man in his twenties would be seen as suffering from a major disorder because of probable poor reproductive success, while an equally 'psychotic' male over 50 years old who had no children would be seen as less severely ill."[91]

The major sociobiological theorists have expressed astonishment, outrage, and genuine anger at the force of expressly political attacks on their work. They had, they pointed out, not the least intention of working against social change. "While I was doing my own work," wrote Trivers, "it never occurred to me that the work was actually serving deeply regressive political aims."[92] And, indeed, I have every reason to take Trivers at his word. For apparently as these scientists developed their theories they never questioned their stance as politically neutral investigators interested only in empirical truth. Whatever science finds, after all, must be openly confronted. The truth is the truth, and if it's unpleasant so be it. Let's not kill the messenger because he bears sorrowful tidings.

But having been forced from the Eden of Scientific Purity into the political fray, the sociobiologists have taken up the challenge. Reserving the scientific mantle for themselves, they attack their critics as ideological, politically expedient, "intellectually feeble, and lazy."[93] Worse yet—they are Marxists (indeed some of them *are* Marxists, although they are also highly respected scientists).

"Marxism and other secular religions offer little more than promises of material welfare and a legislated escape from the consequences of human nature,"[94] writes Wilson, who instead suggests we search for a new morality "based on a more truthful definition of man."[95] In thus claiming to have his finger on the scientific truth, Wilson hopes to discredit his detractors as "metaphysical" ideologues. But in launching the counterattack in this fashion he inevitably takes explicit political positions, contrasting his "analytic program of research on human society" with "the metaphysical truths of Marxism."[96] We have seen, however, that the programs for change proposed by people such as the Shields, Symons, Wilson, and MacKinnon always circle back to a suggestion that things aren't so bad as they stand. Furthermore, rather than confront their critics on scientific grounds, Wilson and others have launched their own political attacks. In taking their stand as scientific Custers opposing the Marxist Indian attack, they have declared themselves defenders of an ideology of scientific rationalism. Yet whether they like it or not, their claims about human nature fall so far short of being scientifically defensible that one can only consider human sociobiology as a political science. It belongs in the arena of philosophical and political controversy far more than in that of scientific debate. The critics of sociobiology make no bones about their beliefs. It is high time that those who make pronouncements about human sociobiology own up to theirs.

7

SEX AND SCIENCE: A CONCLUSION

> The biological argument is different. . . . With it the alleged authority of "science" is placed squarely behind the notion that it is not just politically questionable but practically futile to propose collective social action to eliminate . . . major inequalities that divide us.
>
> —PHILIP GREEN
> *The Pursuit of Inequality,* 1981

CONSIDER the average female. A composite drawing derived from the research findings presented in this book would describe an odd, sometimes incredible creature. She is a skilled housewife; but should she enter the work force, she will remain at the bottom or intermediate levels of the business, government, or professional hierarchy. As a result she will be poor, failing to earn a status worthy of cash rewards because she is not good at math; she may talk too much and too superficially; she is a potential victim; and she is unaggressive except when caught in the throes of some hormonal imbalance which may then turn her to irrational emotional outbursts and even, in extreme cases, to violence. Furthermore, what creative potential she has left over from child rearing peaks relatively early in her life cycle—at about fifty years of age—and thereafter she slowly slides downhill physically, emotionally, and intellectually.

Although the companion portrait of the average male is not entirely complimentary, it is in large part positive and in any case carries with it financial compensation. In general, men have evolved to become "controlled, cunning, cooperative, attractive to the ladies,

205

good with the children, relaxed, tough, eloquent, skillful, knowledgeable and proficient in self-defense and hunting."[1] They are likely to become successful businessmen, politicians, or scholars because of their controlled, unemotional aggressiveness and excellence at mathematics. Still, not all men luck out: a chance throw of the genetic dice causes some to fail in the competition for political power, social dominance, money, and mates. One sorry legacy of this biological fact is that while all men are potential rapists, some carry out that potential.

In this picture society emerges as fair and just. There is no significant wage or job discrimination. Ability determines income distribution; poverty results from individual incompetence. Nor is there anything wrong with the way our educational system works. If women do poorly at math it is because their brains work differently. Under the influence of "raging hormones" women and men, it seems, may both act irrationally—women killing their men friends and beating their children, men raping strangers and beating their wives. The courts are asked to treat these antisocial behaviors as biologically caused illnesses rather than as criminal activities. In interpersonal relationships men who fail to care for their families and women who flirt and play hard-to-get are simply responding to their genetic "deep structure." There is, in short, a lot at stake in using or objecting to the use of biological explanations of human behavior. The fundamental issue is social change—do we need it and is it possible?

I have obviously stated the most extreme composite cases. Yet it is not I who have formulated these ideas. I (along with others)[2] have simply underlined their presence in the scientific literature. With the possible exception of Stephen Goldberg,[3] probably none of the other scientists whose work I have criticized consciously believe that political, social, and financial male dominance is biologically inevitable. Perhaps, then, the metaphor of a composite drawing is itself inaccurate; maybe we should instead envision these scientists as having created a montage while blindfolded, unable to see or acknowledge the overall construction to which they contributed, a sort of scientific pin the tail on the donkey.

But it isn't really. Because in this case the donkey is a human female and the artistic composition and its implications far from random. Further, the implications are highly politicized, at the levels of both interpersonal and societal politics. One cannot help but be astonished at the apparently unself-conscious Everyman's fantasy reflected in sociobiological descriptions of men as cunning, charming, legitimately trying to mate with as many women as possible while becoming powerful businessmen and scholars. I, for one, find the lack of introspection displayed by these (possibly) self-portraits quite astounding. While it is annoying to encounter individuals attempting to live out the fantasy, it is not at the interpersonal level of the "battle between the sexes" that the most widespread damage is done. Society incurs the greatest costs from social policy based on biological views about the origins of political equality, poverty, and equal opportunity and the (im)possibility of social change.

One version of the American feminist vision(s) foresees a world of total equality. Parents would share equally in child care while mates—both hetero- and homosexual—would live in relationships of mutual respect, openness, fidelity, and honesty. In this world of the future men and women would fully share political and financial power; no one would be unable—in the midst of great wealth—to feed and clothe their children adequately. Men and women would be represented equally, according to their equal abilities, in all walks of life.

The composite biological picture considered in this book, however, permits no such world. Women are naturally better mothers, while men are genetically predisposed to be "aggressive, hasty, fickle and indiscriminating."[4] If need be they will rape to pass on their genes, and the extreme measures society would have to take to change these circumstances would in themselves be so repressive as to be unacceptable. Furthermore, women's lack of aggressive drive and native ability ensures that they will always learn less; laws guaranteeing equal pay, ironically, would discriminate against the hardworking, hard-driving, and efficient men in the work force. So too would affirmative action programs, which would simply legislate the hiring of less qualified women. In sum, the biological program for society holds that unless one uses extreme measures of social coercion, more women will stay home while men will predominate in business and government. The increasing slide of women into poverty may not be preventable. This, I need hardly point out, is a very clear social vision, one at political odds with the feminist view of the future.

One central contention of this book is that there is no such thing as apolitical science. Science is a human activity inseparable from the societal atmosphere of its time and place. Scientists, therefore, are influenced—consciously or unconsciously—by the political needs and urgencies of their society. Such needs can be recognized in the use to which the scientific discovery is ultimately put (as with the development of the atomic bomb). In our culture scientific research embodies inequality in the process of scientific production, a system in which there are scientists who get the honors and rewards and assistants who do the work. The most commonplace influence our society exerts on scientific activity is the direct political authority by which Congress can determine what kinds of research and how much of it will be supported.

Good Science, Bad Science, Feminist Science

Throughout this book I have taken a dual tack. I have asked of each claim about women and biology a very conventional scientific question: "What is the evidence?" At the same time I have scrutinized the data with an unconventional, feminist eye. While attempting to discredit the old ways of thinking I have tried to speak for newer, more complicated and at the same time more accurate and interesting ways of thinking about the issues at hand. Put another way, I have framed the issue as poorly done science, while at the same time undertaking to look beyond the existence of good science to something called feminist science. I have thus attempted to transform the conventional opposition of good versus bad science, in which only objective, universally agreed upon facts prevail, into a more complicated analysis of the scientific process.

Stephen Jay Gould writes in a review of Dr. Ruth Bleier's *Science and Gender* that he is:

> ... not convinced that many methodological improvements now slowly making their way within science are, as Dr. Bleier argues, especially feminist ways of thinking—the rejection of dualism, the focus on interaction rather than dominance, the abandonment of reductionism for a holistic vision [are only] ... the taxonomic way of thought that naturalists—most of them men—have been urging against reductionist biology for centuries.[5]

Gould, I think, is only partly correct. If the positive focus of a new

208

way of looking at biology is on complexity, holism, and interaction, then certainly more than one current of thought flows in those directions. That this is true has been brought home to me most clearly by finding, within all of the fields I have criticized, working scientists (often lesser known) who write insightful, often devastating accounts of the research in their fields. On occasion, as with Gould himself, those critics are men.* But often, especially in psychological and hormone biology research, they are women and, not accidentally, feminist scientists who have waged intense battles for the opportunity to do scientific work in the first place.[6] Their very status as outsiders—women and feminists in a masculine scientific world—has lent them a vision which quite appropriately claims the label of feminist. The fact that (as Gould points out) overlapping insights into biology have been achieved by others who traveled to a similar endpoint along alternate roads does not invalidate the feminist label. It merely confirms that more than one road can lead to the same destination.

It may be useful at this point to return to the work of Dr. Randi Koeske, and in particular to her theoretical analysis of menopause research.[7] Reading her work while reflecting on the question of feminist versus good science is a little like staring at one of those prints by Escher which have within them, in dark and light, a series of birds and fish. At first one sees the birds, but after staring a while the birds seem to swoop out of sight and rows of fish swim into view. Both sets of animals actually form part of a united pattern, but at some moments the birds stand out while at others the fish enter one's consciousness. Such is the relationship between good science and feminist science.

Koeske offers us her general views about science, followed by a categorization and critique of previous research on menopause, and finally a preview of the menopause research of the future. She finds unsatisfactory both the two current biomedical and the two current behavioral models. The first biomedical model suggests that all of the "symptoms" of menopause—both psychological and physical—result directly from estrogen withdrawal. In the second biomedical model the psychological aspects of the "change of life" lie at the bottom of a cascade that starts with hot flashes, which in turn cause loss of sleep, followed by depression and inertia. Koeske designates

---

* Similarly, as exemplified by the work of Dr. Katharina Dalton, not all female scientists do feminist science. The category of feminist and masculinist are ideological, not biological.

the two behavioral frameworks as the *premorbid personality* model and the *coincidental stress* approach. While both biomedical models suggest that menopausal symptoms are nigh unto inevitable, the premorbid model suggests that the presence or absence of psychological symptoms depends upon how well-adjusted and stable a woman is *before* menopause. On the other hand, as its label suggests, the coincidental stress model looks especially at all the other things going on in a woman's life during the period when she experiences menopause. Thus career changes, children leaving home, growing evidence of aging, changes in marital patterns, and the death of one's age peers all become focused or symbolized by the event of menopause; the biological changes involved do not in themselves *cause* all of the physical and psychological changes undergone during that period of one's life cycle.

Both the biomedical and the behavioral approaches ignore key aspects of women's existence and, more important perhaps, fail to appreciate the interaction between psychological and physiological states. The medical world, for example, views women as either normal or abnormal and assumes that their pathological reactions to menopause result directly or secondarily from declining estrogen levels. Such an analysis ignores the enormous individual variation in women's experience by categorizing personal descriptions of reality as "subjective" rather than scientific; it also turns a deaf ear to the ways in which women's social realities differ from those of men. On the other hand, most behavioral science models posit an oversimplified, one-to-one connection between social structure and individual experience. Put another way, the behavioral science models usually emphasize the "mind" side, while biomedical models play up the "body" side of the classical mind/body separation. This division of human reality has plagued the analysis of health and disease, both physical and psychological, ever since Descartes articulated his "clock model" of the human organism. (That is, the idea that, if it's broken, take out the faulty gear and put in a new one. This approach is taken by those who do heart, kidney, and liver transplants. The model fails to address the question of why the gear broke in the first place.)

Koeske does not provide a full-dress alternative to the biomedical and behavioral models. Instead she proposes a number of new areas and modes of investigation designed to switch us onto different sets of tracks. Her route, she hopes, will be more appreciative of the diversity of women's experiences, be they biological, psychological, or social. To begin with she suggests developing some insight into the complex nature of body-behavior relationships. She emphasizes that

> . . . rather than a single one-to-one relationship existing between physiological processes and "symptom" reports . . . many such relationships exist. Variations may occur . . . in the number of different physiological processes . . . associated with the same "symptom," the openness of the underlying processes to environmental influence, the accessibility of the involved processes to conscious experience . . . [and the] incorporation of these processes into one or more system of explanation by different social groups.

Furthermore, we must simultaneously acknowledge the interaction between "pure" biological variation and cultural variation, the latter greatly influencing such things as "diet, exercise, obesity, sleep . . . parity, lactation, [and] . . . available medical care," all of which can and do affect the experience of menopause.[8]

In a move lifted straight from the (feminist) women's health movement, Koeske's first practical research suggestion is to take seriously women's own accounts of their experience. Noting that most chronicles of menopausal experience have been written by male doctors observing women who come to them for medical treatment, Koeske emphasizes the need to do more of the sort of widespread data collection done informally by the Boston Women's Health Collective (see chapter 4). Researchers must learn to use the "subjective" reports from women as valid data that provide legitimate insight into our understanding of the menopausal experience. In addition, researchers must learn to view the body and behavior as part of an interconnected system in which many different things happen at many different levels. To focus on only one such level inevitably distorts the picture. Koeske, therefore, urges the development of "studies that trace pathways toward experience up from the level of biochemistry and down from the level of culture or social context."[9] In addition, she urges researchers to figure out— as the work of Goodman et al. (discussed in Chapter 4) tries to do—just what features distinguish menopausal experience from other life experiences. Of course in order to carry out any of these tasks researchers must begin to include psychological and physiological variables in the same study.

Scientists, in Koeske's view, create models of reality into which they insert various things they can observe, directly or indirectly.

In working from such models researchers frame particular questions, decide what are and are not appropriate data, and decide which kinds of controls to use and how to analyze the information collected. Koeske's view of science is, in fact, the same as the one we have developed throughout this book. One implication of this "model of reality" viewpoint is that scientific research is not infinitely "objective." As Koeske writes, science is not "divorced from perspectives so that it becomes unequivocally possible to separate sets of data into the mutually exclusive categories of "error" and "truth." This process of validating truths is only possible . . . within the assumptions of a particular perspective."[10] Rejecting the idea that some methods, techniques, or approaches are inherently more "objective" or "real" than others, she argues instead that all methods have inherent areas of uncertainty; the usefulness of a particular research approach can be judged only in the light of the problem for which it is chosen.

Positively speaking, science, while not providing an absolute reality, can exclude areas of uncertainty by systematically investigating how well particular models of reality stand up under experimental test. Repeated observations, for example, of what happens when I take a book, hold it two feet above the floor, and then let it go, leave me quite certain that it will always fall down rather than float upward. Always, that is, as long as I do the experiment within the pull of the earth's gravity. In this context, gravity is a concept— not a tangible thing—which scientists have named to describe an aspect of our day-to-day experience, an aspect that is more consistent, less variable, and thus more predictable than *any* aspect of human behavior—especially any of the supposed male/female differences discussed in this book.

I began this detailed discussion of Koeske's work because I felt it illustrated the dilemma of trying to distinguish unequivocally between science well done and science that is feminist. At one level Koeske's analysis and suggestions for future work all merit the label of scientific advance. Now that she has shown the way, a better model can emerge. Good science in the long run prevails over bad. Although I do think Koeske's approach is solid science, clearly better and—if you will—more scientific than the work she criticizes, I nevertheless feel that an analysis of good science versus bad science does not accurately represent the complexities of scientific research.

To begin with, it is impossible to ignore the role of expressly feminist insights concerning the subjective/objective separation, the validation of a woman's individual health experiences, the highlighting of the fear and dislike of women frequently found in the medical literature, and the complexity and social contexts of women's lives. Feminist scientists insist that these factors affect women's health and behavior (as well, I hasten to add, as vice versa). These ideas, although they may represent good science, arose in the context of a vast and multiply branched political-cultural movement, that of modern Western feminism. To hold out for a good versus bad science analysis is to ignore the important role feminism has played in *forcing* the re-evaluation of inadequate and often oppressive models of women's health and behavior.

It's not just that we feminists want credit and recognition for the impact we've had. We certainly do (naturally)! But if we don't recognize the approaches developed by scientists such as Koeske as one aspect of broad political change, then a real danger exists that "good science" will *not* prevail. In the past, legions of highly trained doctors and scientists have failed to see and criticize what is wrong with the biomedical and behavioral models of female behavior. Why? Because, I believe, they had no alternate framework within which to develop new sight. Feminism provided that new vision, allowing many scientists—even those who do not consider themselves political feminists—to move in a new direction. "Good science" in the absence of a political and cultural movement did not get very far. And even now, it is an uphill battle. The women and men attempting to redefine biosocial research have not gained the recognition, professional status, or funding which seems so easily to go to the more conventional researchers. Quality research alone is not enough. *Good science*—which in this historical moment incorporates many insights from feminism—can prevail only when the social and political atmosphere offers it space to grow and develop.

The Unbound Foot: Women and Athletics

Throughout this book I have looked at two somewhat different sorts of biological/sociological questions. The first, discussed in some detail earlier, involves aspects of physiology that are special to women. These include menstruation and menopause. Although not discussed herein, physiological and psychological changes associated with childbirth—another uniquely female activity—also fall into this category. Accounts of uniquely female events, however, often intermingle with a second, different type of question, one that looks for and examines male/female differences. The question of male/female differences—just what they are and how great they might be—takes up space in many a cocktail party conversation as well as in many a scientific discussion. I have argued that little credible evidence exists to support the ideas that men and women have different mathematical and verbal abilities; I hold a similar viewpoint on the subject of sex-related differences in aggression. But how far would I take this lack of difference? Surely I must acknowledge that men and women are physically different. Where do I step over the line from denying difference to acknowledging it?

Let me begin with the obvious. Except for the incongruous cases of sexual development discussed in chapter 3, men and women have different reproductive systems and organs. They also have hormones that may differ in amount although not in kind. These differences may be small or large, depending on biological rhythms, life cycle rhythms, stress, and both lifelong and immediate individual experience. On average men are a bit taller and a bit stronger than women. Obviously a physiological, inherent, natural difference, you say? Here I begin to hedge my bets, a hedging that I can perhaps illustrate most usefully by examining male/female differences in athletic performance.

It is easy to forget that only in very recent times have girls been allowed, even as young children, to roam free and to train their bodies. Athenian women had permission neither to watch nor to participate in the original Olympic games; not until 1984 did females officially run in the Olympic marathon. One nineteenth-century American writer considered woman to be a true physical oddity: "The width of her pelvis and the consequent separation . . . of the heads of her thighbones *render even walking difficult*"[11] (emphasis added), while a well-known twentieth-century "sports philosopher" suggested that "one way of dealing with these disparities between the athletic promise and achievements of men and women is to view women as truncated males."[12] Only within the last decade have educators even considered the idea that boys and girls should or could receive the same athletic training. The argument has always been that the biological differences between boys and girls and men and women prevent their competition on the athletic field. Yet one need only read the following poignant account of Chinese foot binding to remember that we still do not know how far a female athlete might have gone had she, from infancy, used her body as fully and as freely as do many little boys:

> They did not begin to bind my feet until I was seven because I loved so much to run and play. . . . [Then] they had to draw the bindings tighter than usual. My feet hurt so much that for two years I had to crawl on my hands and knees. Sometimes at night . . . I could not sleep. I stuck my feet under my mother and she lay on them so they hurt less . . . by the time I was eleven my feet did not hurt and by the time I was thirteen they were finished.[13]

The knowledge that women's feet have only recently been unbound, however, cannot hide the fact that grown men and women look different, that they are sexually dimorphic. Although humans are among the least dimorphic of primates, the fact that the average adult female weighs 10 percent less than the average male (among gorillas the average female weighs 50 percent less than the average male) remains obvious. So too do the variety of other physical characteristics, particularly adult muscle shape and size and the amount and distribution of body fat. Hormones most likely are an important factor in the development of height, muscle size, and fat distribution, although their specific mode of action is a matter of some dispute.[14]

During their early years, girls and boys are quite similar in height. By the early teens girls, who mature earlier but stop growing sooner, spurt ahead, but boys continue to grow for three to five years longer and thus on average reach a taller height. One college athletic recruiter commented that it is easier to recruit female athletes as first-year students because they are a known quantity, whereas the changes in height and strength for boys continue throughout the college years. The key hormone in this growth process is called, appropriately enough, *growth* hormone. The pituitary gland that produces it probably uses a somewhat different biological clock in males than in females, but even this is not certain. Some reports suggest that exercise can affect the short-term synthesis of growth hormone; thus the different activity levels of boys and girls could alter growth hormone metabolism.[15] There are some suggestions that physical training in children leads to more vigorous growth.[16] Even today the amount of physical activity that boys and girls engage in varies greatly, although the differences must certainly be less than they were a hundred years ago. As social restrictions on little girls continue to fall by the wayside, it will be of interest to see whether one result is a decrease in the 10 percent height difference between men and women. My suspicion is that, given widespread changes leading to equal physical activity for boys and girls in an environment with adequate nutrition, the male/female height dimorphism will decrease by a few percentage points but will not disappear altogether.

About 25 percent of the body weight of nonathletic women is fat, in contrast to the 15 percent of the average untrained man's body weight.[17] (The fact that women usually have more of that fat in their breasts and legs is no secret.) This physical difference appears at puberty and, although poorly understood, is thought to result from the greater amount of estrogen active in the female. As anyone who has ever dieted knows, however, body fat is not constant. The amount present in the bodies of highly trained female long-distance runners approaches that of similarly trained men,[18] although what is there is still differently distributed. Some physicians think that the difference in fat content between average college-aged men and women is primarily due to differences in life-style.[19] This, too, only changing social customs will allow us to know for sure.

Height and shape differences are not absolute, but it may be that strength differences are. During development the cells that become muscle fuse with one another to form large fibers. The number of fibers in each individual becomes fixed during the first few years of life, and subsequent muscle growth consists only of increases in the length and width of such fibers. Much of the muscle size differences between males and females result from disparities in fiber growth rather than fiber number. Both hormones and physical activity play a role. Growth hormone differences may account for the fact that girls' muscles grow to their maximum size at an earlier age, while the combination of a more prolonged period of growth hormone synthesis in boys and increases in testosterone level may, together, account for the greater muscle bulk evident as young men mature.

The belief that testosterone builds muscle strength has contributed to a controversy in the sports world. Should athletes take

androgen-like drugs—paying the price in future health problems—
in order to build up their bodies? Although these chemicals do
promote weight gain and increased thigh circumference, controlled
studies show no significant differences in strength between men
who do and do not take these drugs. Ironically, the total amount of
blood testosterone has been shown to decrease in men taking the
externally supplied androgen, a change mediated by the lowered
blood concentrations of a hormone-binding protein.[20] Despite such
studies, belief in the effectiveness of androgenic drugs remains,
continuing to provide serious difficulties for athletes, their physicians, and the organizers of athletic events.

Even for height, body shape, muscle fiber number, and physical
differences in muscle shape, hormones alone tell only a partial story.
This is made clear from an observation of Balinese men recorded
by the anthropologist Margaret Mead.

> The arms of the men are almost as free of heavy muscle as those of
> the women, yet the potentiality for the development of heavy muscle
> is there; when Balinese work as dock-coolies . . . their muscles develop
> and harden. But in their own villages they prefer to carry rather than
> lift, and to summon many hands to every task. . . . *If we knew no*
> *other people than the Balinese we would never guess that men were*
> *so made that they could develop heavy muscles.*[21] [Emphasis added]

The question of muscle strength seems to be somewhat separate
from that of shape. Some estimates suggest that even highly trained
athletes use only 20 percent of their muscle potential, and changes
in strength may well involve not only increases in muscle bulk but,
perhaps as important, changes in the use of muscle that is already
present. Strength is the ability of an individual to exert force against
some external resistance, and different parts of the body have
different strengths. The average strength differences between men
and women result at least in part from men's larger size. The upper
body strength of the average female (that is, strength derived from
arms and shoulders) is about half that of the average male although,
when matched for size, a woman has 80 percent of a man's upper
body strength. The lower body strength of the average woman
reaches 70 percent of the average man's, and when the comparison
is made between individuals of the same weight a woman's lower
body strength approaches 93 percent of a man's. Leg strength
measured relative to lean body weight (leaving out the fat differences)
actually shows women's legs to be 5.8 percent stronger than men's.[22]

One implication of these data is that sports emphasizing upper body strength will probably always offer males an advantage—as long as they are played in a culture such as ours, rather than in a place like Bali. Advantages accruing to men in other sports such as running, however, may be due only to differences in leg length, rather than strength.

In this discussion of differences between untrained men and women, the relative influences of height, fat content, strength of hormones, and environment cannot be easily untangled. At the moment it appears that differences in the timing of growth hormone synthesis during childhood and adolescence may account for male/female height differences and may also be a component of differences in muscle development. On the other hand, it remains possible (and only time will tell) that at least some of the height and strength dimorphism between males and females would diminish in a culture in which girls from infancy on engaged in the same amount and kind of physical activity as boys. It is my own guess, though, that even then small average differences would remain. Finally, it behooves us to remember that the amount of variation among men and among women is greater than that between the sexes. Thus no two differently sexed individuals can be assumed, sight unseen, to have different heights, shapes, or strengths.

Looking at highly trained athletes offers another view of physical differences between men and women. Although one assumes that both groups will compare more closely because they have had a greater chance through training to develop their potential, few female athletes begin to train as early or have the same opportunities and training as do male athletes. One sport in which differences in body composition redound to women's advantage is marathon swimming, women's greater body fat providing increased buoyancy and protection against the cold. Here women hold the world record, a title that came easily once it became acceptable for females to try the feat. The first woman ever to swim the English Channel (Gertrude Ederle in 1926) not only astounded the world by succeeding at all, but she broke the men's record by two hours! In fact she so took the public by surprise that a London newspaper did not have time to withdraw an editorial claiming that her failure (expected when the paper went to press) demonstrated that women were physically inferior to men and their entry into competitive athletics a hopeless enterprise.[23]

Between 1964 and 1984 women marathon runners have knocked

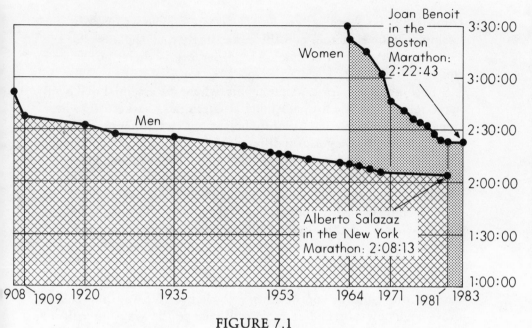

FIGURE 7.1

History of the World Record in the Marathon

NOTE: Jane Gross, "Women Athletes Topple Sports Myths," *New York Times*, 12 Aug. 1984, p. 22E. Reprinted by permission of the Amateur Athletic Union of USA. The marathon is 26 miles, 385 yards long. Finishing times are given in hours, minutes, and seconds.

more than an hour-and-a-half off their running times, while men's times during that same period have decreased by only a few minutes (see figure 7.1). The relative differences between men's and women's times in shorter running events have also fallen considerably since the 1930s. In 1934, for example, the women's time for the 100-meter run was 13.5 percent lower than the men's, but by 1974 the difference had decreased to 9.1 percent. There are similar trends in swimming (with the possible exception of the 100-meter event). In all swimming and track events in the 1976 Olympics, females were 89 to 93 percent as fast as men—that is, females were 10 percent slower.[24] If the gap between highly trained male and female athletes were to continue to close at the current rate, in thirty to forty years men and women would compete in these sports on an equal basis.[25] It is, of course, also possible that the rate of female improvement will level off. In that case we will have a better idea of just how different the physical capacities of the male and female body are. One way to guess about the outcome is to look at male and female differences in a country where training and coaching methods seem

219

comparable for both sexes. East German female swimmers, for example, swim a mere 3 percent more slowly than the men (for the 100- and 400-meter freestyle). The difference is there, but it isn't much!

In some sports, for example tennis, where the essential overhand serve relies heavily on upper body strength, and basketball, where upper body strength and absolute height are vital, men and women will probably always perform differently. Others, such as gymnastics, may well turn out to favor women. Whatever the outcome, however, it has become clear that girls and women can be excellent athletes, and it has become increasingly acceptable for the average girl to learn to enjoy using her body in physical activity. There are hormonal bases for some of the physical differences between adult men and women. Yet even these interact with culture and socialization to produce the final product. No matter how our ideas about male and female physique evolve in the coming years, one thing remains certain: our cultural conceptions will change the way our bodies grow, and how our bodies grow will change the way our culture views them.

A Program for the Future

This book is incomplete. I have chosen not to address myself to some aspects of allegedly gender-specific human behavior for which scientists have offered a biological explanation. The areas left unexplored include writings about women and depression, mothering (but not fathering) instincts, and theories about biological "causes" of homosexuality and transsexuality. That I have omitted these areas may violate some sense of completeness. In fact each could be subjected to the same sort of analysis used for the topics I have dealt with. But because I fully believe that the same combination of inadequate research and inappropriate model building would turn up (as a number of shorter analyses of these subjects have shown),[26] the general lesson emerges clearly without having to dwell on every available example. Any biological theory about human behavior that ignores the complex of forces affecting behavior as well as the profound two-way interactions between mind and body is scientifically hopeless. Yet the continued appearance of such ideas in both the scientific and popular literature attests to their tenacity. Their often unself-conscious coupling with (usually retrogressive) social programs helps to explain their continued presence and to underline the political nature of their formulation.

As should be clear by now, I do not argue for a program of behavioral research that ignores biology. Instead I put forth a plea to release biology from its sacrosanct status as First Cause and give it a more appropriate place in the network of disciplines that constitute the proper study of humankind. The challenge—and a hard one it is—is to develop a new model. To do so we will have to commit ourselves to figuring out *how* to do contextual research. We will also have to develop new analytical frameworks, at first taking the sorts of tentative steps indicated by Randi Koeske and others and with time, thought, and more research, allowing our immature models to grow into adult ones. For scientists trained in an ideology of control this is a hard task. To begin with we will have to give up on the idea that the goal of behavioral research is to control behavior, accepting instead the fact that we will never reliably be able to predict human behavior. Instead of striving to be mechanists in biological clothing, we biological scientists must accept the idea that our enterprise is a rather different one, not *softer* than physics but a great deal more difficult. We can take pride in that difficulty and complexity rather than trying to simplify to the point of meaninglessness the phenomena we study.

I do *not*, however, believe that *all* of the research areas covered in this book warrant further investment of time, money, and talent. We need to learn more about menstruation and menopause, especially about those cases for which health problems really do arise. On the other hand, further research into sex differences in cognition or brain laterality seems to me uncalled for. If there are any differences at all, they are so small that intrasexual variation swamps them. The real issues in cognition center upon how one learns (both in and out of the classroom). As Dr. Sheila Tobias has recently stated the problem:

> The issues ... have less to do with excellence than with resource allocation; less to do with education than with our nation's skewed investment in weapons science; less to do with assessing and punishing teachers than with re-establishing substantial support of math-science teaching ... and less, much less, to do with re-establishing putative genetic inferiority of women and minorities than with establishing equal, really equal opportunity for young people to learn science.[27]

221

Similarly, it is not worthwhile to continue to do research on animal aggression in highly artificial settings. The trend toward understanding community and context in animal behavior research, on the other hand, is encouraging. Still, as exemplified by the work on rape, to take even very extensive animal research, define it according to uniquely human behaviors, and then to use it to analyze human behavior is both logically flawed and politically dangerous. The challenge to the scientific community, then, is to clean up its act and to undertake a new program of research which emphasizes complexity, mind-body interactions, and human flexibility.

That challenge, of course, cannot be met in the absence of a broader social program. Only in a society fully committed to educational equity can we develop research programs that focus on the learning and teaching process rather than on the possibility of inherent racial or sexual deficiencies. Only in a culture fully committed to economic and social equality can we have research programs that examine behavior in the context of present and future possibility rather than permanent limitation. And only in a culture that genuinely respects and values members of both sexes will respectful and healthful analyses of both female and male medical problems emerge, analyses that recognize reproductive differences as normal and that can make well-informed distinctions between healthy physiological activities and disabling states of disease. The two challenges—scientific and social—are profoundly interconnected. Neither the challenge to scientists to construct a more scientific and—yes—a more feminist research program, nor the challenge to all of us as world citizens to build a society that respects and recognizes difference while understanding and emphasizing human similarities, can be met separately.

APPENDIX

Approaches to Measuring Fitness

The relative probability of survival and rate of reproduction of individuals of a particular genotype is a statistical concept. Any particular individual may meet with an untimely end, even if he or she is vigorous and potentially quite fertile. Because a similar individual might in fact reproduce quite successfully, the chance loss of life does not mean the individual had an unfit genotype. Thus the relative fitness of a particular genotype can be measured only from a group average, never from the life history of any one individual. Identical twins provide an example, as they both have exactly the same genes. Suppose, though, that one twin was killed by lightning at the age of ten while the other survived to adulthood and had children. Because their genotypes were identical the survivor and his or her twin, in the Darwinian sense, are equally fit. Yet one reproduced and the other did not.

According to the Modern Synthesis, individuals of some genotypes are fitter on average than those of others, and thus become more frequently represented in the population. Evolution occurs as the gene frequencies of a population change. The problem thus stated seems simple. In order to study evolution in action, all one needs to do is *measure* the fitness of a particular genotype and observe how it comes to dominate a population, or how the fitness changes in different environments. It ought to be possible to study such matters both in the field and by using "artificial" laboratory evolution. Nevertheless, in 1974 a top evolutionary geneticist wrote: "To the present moment no one has succeeded in measuring with any accuracy the net fitness of genotypes for any locus in any species in any environment in nature."[1] The problem is practical. Two measurements are required— the total fecundity of individuals of a particular genotype and the probability that a newly formed egg of a certain type will survive to reproductive age (fertility and survival). A female's fertility is measured by classifying her genotype and counting how many eggs she lays. Measuring a male's fertility is not quite so easy. In completely monogamous organisms one can infer a male's fertility from the number of offspring his mate produces, otherwise one has to analyze the offspring to see which males contributed to their production. This in turn requires knowing enough about the female parent to separate out the maternal and paternal genetic contributions.

Even knowing all that, the problem still remains of figuring out how likely it is for an embryo with a particular genetic makeup to survive— knowledge virtually impossible to obtain because one infers genetic makeup

223

from developed characteristics, not from embryos. One does not, for example, know if a cat embryo carries the gene for white or black coat color until the kitten is born. One can use adult genotype frequencies to get some idea of survival, but in natural populations such frequencies reflect not only the probability of survival to adulthood, but also the immigration and emigration into and out of a population.

Faced with such difficulties, experimentalists have resorted to measuring components of fitness. The fruit fly *Drosophila* has myriad genetic variations which evolutionists have studied in the laboratory, where it is possible to see how many eggs survive to adulthood. For instance, one could study eggs laid by "pure" (homozygous) white-eyed and "pure" red-eyed flies. If on average 95 percent of eggs from red-eyed flies hatched and developed into adults, compared with only 75 percent of eggs from white-eyed flies, then one could say that the partial fitness (survival) of the red variant is greater than the white. At least this is theoretically possible. In practice, one must be sure that no unidentified genetic factors interfere with the results[2] and realize that constructing two strains of flies, identical except for one set of alleles, is profoundly difficult, even given the sophisticated state of *Drosophila* genetics.

Ideally, figuring out the average reproductive success of individuals with a particular genetic makeup ought to give one a partial estimate of fitness. In order to get such a measurement one would have to identify animals of particular genotypes and count how many offspring they produce over a lifetime. In the best of all possible worlds one would also need to know something about the observed animal's mate, since a genetically fit individual may produce few offspring if mated with an unfit partner. If one compares the average fecundity of individuals of a particular genotype with those of a differing genetic makeup, one must assume that individuals in the two groups choose their mates at random, so that differences due to mate selection cancel out. While such an assumption is necessary, there is no reason to think that it is really correct. (Still, without it, one might as well fold up shop.)

Quite recently ecologists studying the reproductive success of animals in the wild have started to make lifelong measurements on individuals they know by sight. A group of Scottish biologists, for example, has just completed a twelve-year study of a population of deer living on the island of Rhum. By learning to recognize each individual by sight and by weighing and tagging fawns at birth, they kept track of the reproductive success of all the grown males and females in the population for a full generation. Their heroic work yields interesting results, but it too leaves the question of fitness up in the air, because such a study does not reveal the genetic basis of the differences in reproductive success that were measured.[3]

NOTES

Chapter 1

1. W. Bagehot, "Biology and 'Woman's' Rights," *Popular Science Monthly* 14(1879):201–13.
2. M. K. Sedgewick, "Some Scientific Aspects of the Woman Suffrage Question," *Gunton's Magazine* 20(1901):333–44; G. Ferrero, "The Problem of Woman, From a Bio-Sociological Point of View," *The Monist* 4(1894):261–74; E. D. Cope, "The Relationship of the Sexes to Government," *Popular Science Monthly* 33(1888):261–74; J. Weir, Jr., "The Effect of Female Suffrage on Posterity," *American Naturalist* 29(1895):815–25.
3. H. Maudsley, "Sex in Mind and in Education," *Popular Science Monthly* 5(1874):198–215; A. L. Smith, "Higher Education of Women and Race Suicide," *Popular Science Monthly* 66(1905):466–73; G. DeLawney, "Equality and Inequality in Sex," *Popular Science Monthly* 20(1881): 184–92; M. A. Hardaker, "Science and the Woman Question," *Popular Science Monthly* 20(1882):577–84; N. Morais, "A Reply to Miss Hardaker on: The Woman Question," *Popular Science Monthly* 21(1882):70–78; Stephen J. Gould, *The Mismeasure of Man* (New York: Norton, 1981); W. L. Distant, "On the Mental Differences Between the Sexes," *Journal of the Anthropological Institute* 4(1875):78–85.
4. Janet Sayers, *Biological Politics: Feminist and Anti-feminist Perspectives* (London: Tavistock, 1982).
5. Antoinette B. Blackwell, *The Sexes Throughout Nature* (New York: Putnam, 1875; Westport, Conn.: Hyperion, 1976).
6. E. B. Gamble, *Evolution of Woman: An Inquiry into the Dogma of Her Inferiority to Man* (New York: Putnam, 1893; Westport, Conn.: Hyperion, 1976).
7. D. Barash, *The Whisperings Within* (New York: Penguin, 1979); D. Symons, *The Evolution of Human Sexuality* (New York: Oxford University Press, 1979); and J. Durden-Smith and D. DeSimone, "The Main Event," *Playboy*, June 1982, 165.
8. Symons, *Evolution of Human Sexuality.*
9. Katharina Dalton, *The Premenstrual Syndrome* (London: William Heinemann Medical Books, 1964), 94.
10. M. B. Rosenthal, "Insights into the Premenstrual Syndrome," *Physician and Patient* (April 1983):46–53.
11. T. O. Marsh, *Roots of Crime* (Newton, N.J.: Nellen, 1981).
12. W. Herbert, "Premenstrual Changes," *Science News* 122(1982):380–81.
13. D. E. H. Russell and N. Howell, "The Prevalence of Rape in the United States Revisited," *Signs* 8(1983):688–95.
14. California Commission on the Status of Women, *California Women* (June–July 1983):1.
15. George Gilder, *Wealth and Poverty* (New York: Basic Books, 1981), 137.
16. Michael Levin, "The Feminist Mystique," *Commentary* (Dec. 1980):25–30.
17. U.S. Department of Labor, *Occupational Outlook Quarterly* (Spring 1982): 1–34.
18. J. Sherman, *Sex-related Cognitive Differences* (Springfield, Ill.: Charles C Thomas, 1978); and C. P. Benbow and J. C. Stanley, "Sex Differences in Mathematical Ability: Fact or Artifact?," *Science* 210(1980):1262–64.
19. Advertisement in *The Providence Journal*, 17 June 1983.
20. Marsh, *Roots of Crime.*
21. Sayers, *Biological Politics.*
22. C. N. Jacklin, "Methodological Issues in the Study of Sex-related Differences," *Developmental Review* 1(1981):266–73.
23. S. S. Mosedale, "Science Corrupted: Victorian Biologists Consider 'The Woman Question,'" *Journal of the History of Biology* 11(1978):1–55.
24. T. S. Kuhn, *The Structure of Scientific Revolutions* (Chicago: University of Chicago Press, 1962).

225

Chapter 2

1. "Is Math Ability Affected by Hormones?," *Boston Globe*, 26 Dec. 1983, Science section.

2. Letter to the Editor, *Providence Journal*, 25 March 1984.

3. Margaret Rossiter, *Women Scientists in America: Struggles and Strategies to 1940* (Baltimore: Johns Hopkins University Press, 1982), 107.

4. Maxine Seller, "G. Stanley Hall and Edward Thorndike on the Education of Women: Theory and Policy in the Progressive Era," *Educational Studies* 11(1981): 365–74.

5. Stephanie A. Shields, "Functionalism, Darwinism, and the Psychology of Women: A Study in Social Myth," *American Psychologist* 30(1975):739–54.

6. Rossiter, *Women Scientists;* Seller, "G. Stanley Hall and Edward Thorndike"; and Mary Brown Parlee, "The Sexes Under Scrutiny: From Old Biases to New Theories," *Psychology Today*, (Nov. 1978), 64–69.

7. Seller, "G. Stanley Hall and Edward Thorndike."

8. Rossiter, *Women Scientists.*

9. Seller, "G. Stanley Hall and Edward Thorndike." Seller writes the following: Although the total number of women in the professions increased in the 1920s they remained clustered in "female" occupations. In New York, for example, there were 63,637 (female) teachers and about 22,000 nurses but only 60 Certified Public Accountants, 11 engineers and 8 inventors. There were fewer women in medical schools in the late 1920s than there had been at the turn of the century. [p. 371]

10. Ibid., 273.

11. Ibid.

12. Parlee, "The Sexes Under Scrutiny."

13. Robert G. Lehrke, "A Theory of X-linkage of Major Intellectual Traits," *American Journal of Mental Deficiency* 76(1972):611–19.

14. Anne Anastasi, "Four Hypotheses with a Dearth of Data: Response to Lehrke's 'A Theory of X-linkage of Major Intellectual Traits,'" *American Journal of Mental Deficiency* 76(1972):620–22; Walter E. Nance and Eric Engel, "One X and Four Hypotheses: Response to Lehrke's 'A Theory of X-linkage of Major Intellectual Traits,'" *American Journal of Mental Deficiency* 76(1972):623–25.

15. Nance and Engel, "One X and Four Hypotheses"; Michele A. Wittig, "Sex Differences in Intellectual Functioning: How Much of a Difference do Genes Make?," *Sex Roles* 2(1976):63–74.

16. Robert G. Lehrke, "Sex Linkage: A Biological Basis for Greater Male Variability in Intelligence," in *Human Variation: The Biopsychology of Age, Race, and Sex*, ed. R. Travis Osborne, Clyde E. Noble, and Nathaniel Weyl (New York: Academic Press, 1978), 193.

17. Lehrke makes his argument appear a little more sophisticated from the geneticist's viewpoint by including a discussion of the Lyon Hypothesis, according to which in females the paternally contributed X chromosome in some cells is genetically inactive while in other cells the maternally contributed X is inactive. According to Lehrke this would lead to an "average" expression of X-linked traits, rather than to the extreme he expects in males.

18. Stephen Jay Gould, *The Mismeasure of Man* (New York: Norton, 1981); and Richard C. Lewontin, Steven Rose, and Leon Kamin, *Not in Our Genes* (New York: Random House, 1984).

19. Lehrke, "A Theory of X-linkage."

20. Robert M. Murphy, "Phenyl Ketonuria (PKU) and the Single Gene," *Behavior Genetics* 13(1983):141–57.

21. Wittig, "Sex Differences."

22. Julia Sherman, "Effects of Biological Factors on Sex-related Differences in Mathematics Achievement," in *Women and Mathematics, Research Perspectives for Change*, no. 8, NIE Papers in Education and Work (Washington, D.C.: National Institute of Education, 1977).

23. Lehrke, "A Theory of X-linkage."

24. Anastasi, "Four Hypotheses."

25. Robert Lehrke, "Response to Dr. Anastasi and to the Drs. Nance and Engel," *American Journal of Mental Deficiency* 76(1972):630.

26. Nance and Engel, "One X and Four Hypotheses."

27. Lehrke, "Sex Linkage," 172.

28. Lehrke, "Response to Dr. Anastasi," 631.

29. Osborne, Noble, and Weyl, *Human Variation*.

30. In the epilogue to Osborne, Noble, and Weyl's *Human Variation*, noted geneticist C. D. Darlington writes the following:

> When the Negroids were taken away from Africa by slave traders they were partly . . . rescued from the diseases that had infested them. But slavery did not rescue them from their genetic responses to disease, which have continued with them as an obstacle to their development individually, racially, and culturally. Nor did it rescue them from their primitive beliefs. These experiences teach us indeed that peoples cannot be separated from their history. They are bound down by their evolutionary antecedents, occupational, social and medical. . . . When the Negroids were liberated from agricultural slavery, they were thrown free to shift for themselves in large urban Caucasoid societies. . . . These simple unskilled rural people were suddenly offered irregular urban employment combined with the opportunities of drink and drugs, gambling and prostitution. . . . The intellectually well-endowed races, classes and societies have a moral responsibility for the problems of race mixture. . . . They may hope to escape from these responsibilities by claiming an intellectual and therefore moral equality between all races, classes and societies. But the chapters of this book, step by step, deprive us of the scientific and historical evidence that might support such a comfortable illusion. [pp. 383–84]

31. Barry Mehler, "The New Eugenics," *Science for the People* 15(1983): 18–23.

32. Josef E. Garai and Amram Scheinfeld, "Sex Differences in Mental and Behavioral Traits," *Genetic Psychology Monographs* 77(1968):256.

33. Sandra Witelson, "Sex and the Single Hemisphere: Specialization of the Right Hemisphere for Spatial Processing," *Science* 193(1976):425–27.

34. Daniel Goleman, "Special Abilities of the Sexes: Do they Begin in the Brain?," *Psychology Today*, Nov. 1978, 59.

35. Carole Leland, ed., "Men and Women Learning Together: A Study of College Students in the Late '70s" (Report of the Brown Project, Brown University, 1980).

36. Thorough documentation of the inequality is offered by Rossiter in *Women Scientists*.

37. Garai and Scheinfeld, "Sex Differences in Mental and Behavioral Traits," 275–76.

38. Rossiter, *Women Scientists*, 106.

39. Eleanor E. Maccoby and Carol N. Jacklin, *The Psychology of Sex Differences* (Stanford, Calif.: Stanford University Press, 1974).

40. Julia A. Sherman, *Sex-related Cognitive Differences: An Essay on Theory and Evidence* (Springfield, Ill.: Charles C Thomas, 1978).

41. Anne C. Petersen and Michele A. Wittig, "Sex-related Differences in Cognitive Functioning: An Overview," in *Sex-related Differences in Cognitive Function: Developmental Issues*, ed. M. A. Wittig and A. C. Petersen (New York: Academic Press, 1979).

42. Sherman, *Cognitive Differences*.

43. Carol N. Jacklin, "Epilogue" in *Sex-related Differences in Cognitive Function*, ed. Wittig and Petersen; Janet S. Hyde, "How Large Are Cognitive Differences? A Meta-analysis Using Ω and d," *American Psychologist* 36(1981):892–901; Robert Plomin, and Terryl Foch, "Sex Differences and Individual Differences," *Child Development* 52(1981):383–85.

44. Technically, one calculates the variance as the square of the standard deviation. It is a less intuitively obvious number than the standard deviation but is

a direct measure of how a population varies about a particular mean. Meta-analysis is a method for analyzing the degree to which different variables in a population contribute to the overall variability of the population.

45. Hyde, "How Large Are Cognitive Differences?"

46. Plomin and Foch, "Sex Differences and Individual Differences," 384.

47. Jerre Levy, as quoted in Jo Durden-Smith, "Male and Female—Why," *Quest* 4(1980):17.

48. Maccoby and Jacklin, *The Psychology of Sex Differences.*

49. Sherman, *Sex-related Cognitive Differences.*

50. Julia Sherman, letter to author, 29 April 1981.

51. Hyde, "How Large Are Cognitive Differences?"

52. Ibid.

53. J. M. Connor, M. Schackman, and L. Serbin, "Sex-related Differences in Response to Practice on a Visual-Spatial Test and Generalization to a Related Test," *Child Development* 49(1978):24–29.

54. S. Johnson, "Effects of Practice and Training in Spatial Skills on Sex-related Differences in Performance on Embedded Figures," in Sherman, *Sex-related Cognitive Differences.*

55. E. Fennema and Julia Sherman, "Sex-related Differences in Mathematics, Achievement, Spatial Visualization and Affective Factors," *American Educational Research Journal* 14(1977):51–71.

56. S. B. Nerlove, R. H. Monroe, and R. I. Monroe, "Effect of Environmental Experience on Spatial Ability: A Replication," *Journal of Social Psychology* 84(1971): 3–10.

57. J. W. Berry, "Temne and Eskimo Perceptual Skills," *International Journal of Psychology* 1(1966):207–29; and R. MacArthur, "Sex Differences in Field Independence for the Eskimo," *International Journal of Psychology* 2(1967):139–40.

58. J. W. Berry, "Ecological and Cultural Factors in Spatial Perception Development," *Canadian Journal of Behavioral Science* 3(1971):324–36.

59. H. A. Moss, "Early Sex Differences and Mother-Infant Interaction," in *Sex Differences in Behavior,* ed. R. C. Friedman, R. H. Richart, and R. L. Van de Wiele (New York: Wiley, 1974), 149–63.

60. Stephen Jay Gould, "Women's Brains," in *The Panda's Thumb* (New York: Norton, 1980), 152–59.

61. Shields, "Functionalism, Darwinism, and the Psychology of Women," 471.

62. Ibid.; G. T. W. Patrick, "The Psychology of Women," *Popular Science Monthly* 47(1895):209–25; Franklin Mall, "On several anatomical characters of the human brain said to vary according to race and sex, with especial reference to the weight of the frontal lobe." *American Journal of Anatomy* 9(1909):1–32.

63. Stafford, "Sex Differences in Spatial Visualization as Evidence of Sex-linked Inheritance," *Perceptual and Motor Skills* 13(1961):428.

64. R. P. Corley et al., "Familial Resemblance for the Identical Blocks Test of Spatial Ability: No Evidence of X Linkage," *Behavior Genetics* 10(1980):211–15.

65. Hogben Thomas, "Familial Correctional Analyses, Sex Differences and the X-linked Gene Hypothesis," *Psychological Bulletin* 93(1982):427–40.

66. D. R. Bock and D. Kolakowski, "Further Evidence of Sex-linked Major-Gene Influence on Human Spatial Visualizing Ability," *American Journal of Human Genetics* 25(1973):1–14.

67. Sherman, *Sex-related Cognitive Differences.*

68. Bock and Kolakowski, "Further Evidence of Sex-linked Major-Gene Influence."

69. Michael Levin, "The Feminist Mystique," *Commentary,* Dec. 1980, 25–30.

70. Roger Sperry, "Some Effects of Disconnecting the Cerebral Hemispheres," *Science* 217(1982):1223–26.

71. S. F. Walker, "Lateralization of Functions in the Vertebrate Brain: A Review," *British Journal of Psychology* 71(1980):329–67.

72. Roger Lewin, "Is Your Brain Really Necessary?," *Science* 210(1980): 1232–34.

73. Sperry, "Some Effects of Disconnecting the Cerebral Hemispheres," 1224.

74. Ibid., 1223–26.

228

75. Walker, "Lateralization of Functions," 1224; and R. Restak, *The Brain, the Last Frontier* (New York: Warner Books, 1979).

76. A. Buffery and J. Gray, "Sex Differences in the Development of Spatial and Linguistic Skills," in *Gender Differences: Their Ontogeny and Significance*, ed. C. Ounsted and D. C. Taylor (London: Chirhill Livingston, 1972).

77. Jerre Levy, "Lateral Specialization of the Human Brain: Behavioral Manifestation and Possible Evolutionary Basis," in *The Biology of Behavior*, ed. J. Kiger (Eugene: University of Oregon Press, 1972).

78. D. Goleman, "Special Abilities of the Sexes: Do They Begin in the Brain?," *Psychology Today*, Nov. 1978, 48–59; Tim Hackler, "Women vs. Men: Are They Born Different?," *Mainliner*, May 1980, 122–26; and Jo Durden-Smith, "Men, Women and the Brain: Are Our Brains as Different as Our Bodies?," *Quest/80* 4(1980):15–19, 93–99.

79. Sherman, *Sex-related Cognitive Differences*; J. C. Marshall, "Some Problems and Paradoxes Associated with Recent Accounts of Hemispheric Specialization," *Neuropsychologia* 11(1973):463–70.

80. Levy, "Lateral Specialization"; and Meredith Kimball, "Women and Science: A Critique of Biological Theories," *International Journal of Women's Studies* 4(1981): 318–38.

81. Jerre Levy, "Possible Basis for the Evolution of Lateral Specialization of the Human Brain," *Nature* 224(1969):614–15.

82. Kimball, "Women and Science."

83. Andrew Kersetz, "Sex Distribution in Aphasia," *The Behavioral and Brain Sciences* 5(1982):310.

84. Sherman, *Sex-related Cognitive Differences*.

85. Kimball, "Women and Science"; Kersetz, "Sex Distribution and Aphasia"; and J. McGlone, "Faulty Logic Fuels Controversy," *The Brain and Behavioral Sciences* 5(1982):312–14.

86. McGlone, Jeanette, "Sex Differences in Human Brain Asymmetry: A Critical Survey," *The Behavioral and Brain Sciences* 3(1980):226.

87. Kimball, "Women and Science"; McGlone, "Faulty Logic Fuels Controversy"; and M. P. Bryden, "Evidence of Sex-related Differences in Cerebral Organization," in *Sex-related Differences in Cognitive Functioning*, ed. Wittig and Petersen.

88. Bryden, "Evidence of Sex-related Differences," 137–38.

89. Sherman, *Sex-related Cognitive Differences*; Kersetz, "Sex Distribution in Aphasia"; McGlone, "Sex Differences in Human Brain Asymmetry"; and Bryden, "Evidence of Sex-related Differences."

90. Kimball, "Women and Science."

91. Sherman, *Sex-related Cognitive Differences*.

92. Deborah Waber, "Cognitive Abilities and Sex-related Variations in the Maturation of Cerebral Cortical Functions," in *Sex-related Differences in Cognitive Functioning*, ed. Wittig and Petersen.

93. Nora Newcombe and Mary M. Bandure, "Effect of Age at Puberty on Spatial Ability in Girls: A Question of Mechanism," *Developmental Psychology* 19(1983):215–24.

94. "Are Boys Better at Math?," *New York Times*, 7 Dec. 1980, p. 102.

95. Camilla P. Benbow and Julian C. Stanley, "Sex Differences in Mathematical Ability: Fact or Artifact?" *Science* 210(1980):1262–64.

96. "The Gender Factor in Math," *Time*, 15 Dec. 1980, p. 57; D. A. Williams and P. King, "Do Males have a Math Gene?," *Newsweek*, 15 Dec. 1980, p. 73.

97. *Chronicle of Higher Education* 27(1983):1.

98. Lucy W. Sells, "The Mathematics Filter and the Education of Women and Minorities," in *Women and the Mathematical Mystique*, ed. Lynn H. Fox, Linda Brody, and Dianne Tobin (Baltimore: Johns Hopkins University Press, 1980).

99. Elizabeth K. Stage and Robert Karplus, "Mathematical Ability: Is Sex a Factor?," *Science* 212(1981):114.

100. Sells, "The Mathematics Filter."

101. Elizabeth Fennema, "Sex-related Differences in Mathematics Achievement: Where and Why," in *Women and the Mathematical Mystique*, ed. Fox, Brody and Tobin.

102. Some people argue that it is misleading to compare boys and girls only if they have taken the same number of math courses, because the girls taking large numbers of math courses are a more highly selected sample than the boys. It might be that comparing boys and girls with the same number of math courses really means that one compares bright girls with more average boys. However, Fox, Brody, and Tobin report differential course taking even among highly mathematically gifted girls and boys, suggesting that factors other than mathematical talent are at play when girls and boys decide to proceed into upper-level math courses in high school. See Lynn H. Fox, Linda Brody, and Dianne Tobin, eds., *Women and the Mathematical Mystique* (Baltimore: Johns Hopkins University Press, 1980); see especially the chapter by Dianne Tobin and Lynn H. Fox, "Career Interests and Career Education: A Key to Change," 179–191, in *Women and the Mathematical Mystique*, ed. Fox, Brody, and Tobin.

103. Benbow and Stanley, "Sex Differences in Mathematical Ability," 1263, 1264.

104. Camilla Benbow and Julian Stanley, "Sex Differences in Mathematical Reasoning Ability: More Facts," *Science* 222(1983):1029–31.

105. A. T. Schafer and M. W. Gray, "Sex and Mathematics," *Science* 211(1981): 231. Stage and Karplus, "Mathematical Ability"; Camilla P. Benbow and Julian C. Stanley, "Mathematical Ability: Is Sex a Factor?," *Science* 212(1981):118–21.

106. Benbow and Stanley, "Mathematical Ability: Is Sex a Factor?," 121.

107. Gaea Leinhardt, A. M. Seewald, and Mary Engel, "Learning What's Taught: Sex Differences in Instruction," *Journal of Educational Psychology* 71(1979):432–39.

108. Camilla Benbow and Julian Stanley, "Consequences in High School and College of Sex Differences in Mathematical Reasoning Ability: A Longitudinal Perspective," *American Educational Research Journal* 19(1982):612.

109. C. Benbow, Reply to J. Beckwith and M. Woodruff, "Achievement in Mathematics," *Science* 223(1984):1248.

110. Benbow and Stanley, "Consequences in High School and College," 612.

111. Helen Astin, "Sex Differences in Mathematical and Scientific Precocity," in *Mathematical Talent: Discovery, Description and Development*, ed. J. C. Stanley, D. P. Keating, and Lynn H. Fox (Baltimore: Johns Hopkins University Press, 1974), 82.

112. Lynn Fox and Sanford J. Cohn, "Sex Differences in the Development of Precocious Mathematical Talent," in *Women and the Mathematical Mystique*, ed. Fox, Brody, and Tobin; L. H. Fox, "Sex Differences Among the Mathematically Precocious," *Science* 224(1984):1291–93.

113. Lynn Fox, "The Problem of Women and Mathematics" (Report to the Ford Foundation, New York, 1981); Jayne E. Stake and Charles R. Granger, "Same-Sex and Opposite-Sex Teacher Model Influences on Science Career Commitment among High School Students," *Journal of Educational Psychology* 70(1978):180–86; Doris R. Entwistle and David P. Baker, "Gender and Young Children's Expectations for Performance in Arithmetic," *Developmental Psychology* 19(1983):200–209; Edith H. Luchins, "Sex Differences in Mathematics: How NOT to deal with them," *American Mathematical Monthly* 86(1979):161–68; and Sally L. Hacker, "Mathematization of Engineering: Limits on Women and the Field," in *Machina Ex Dea*, ed. Joan Rothschild (New York: Pergamon, 1983).

Chapter 3

1. Lionel Tiger, "The Possible Biological Origins of Sexual Discrimination," *Impact of Science on Society* 20(1970):37.

2. Maya Pines, "Behavior and Heredity: Links for Specific Traits Are Growing Stronger," *New York Times*, 29 June 1982, section C.

3. S. Mednick, W. F. Gabrielli, and B. Hutchings, "Genetic Influences in Criminal Convictions," *Science* 224(1984):891–93.

4. John Maddox, "Genetics and Heritable I.Q.," *Nature* 309(1984):579; T. W. Teasdale and D. R. Owen, "Heredity and Familial Environment in Intelligence and Educational Level—A Sibling Study," *Nature* 309(1984):620–22.

5. Richard C. Lewontin, Steven Rose, and Leon Kamin, *Not in Our Genes* (New York: Random House, 1984).

6. Philip Kitcher, "Genes," *British Journal of the Philosophy of Science* 33(1982): 337–59.

7. G. Ledyard Stebbins, "Modal Themes: A New Framework for Evolutionary Syntheses," in *Perspectives on Evolution*, ed. R. Milkman (Sunderland, Mass.: Sinauer, 1982).

8. Kitcher, "Genes."

9. Stebbins, "Modal Themes," 4.

10. Stanley Robbins and R. S. Cotran, *Pathologic Basis of Disease* (Philadelpia: W.B. Saunders, 1979).

11. R. D. Alexander, *Darwinism and Human Affairs* (Seattle: University of Washington Press, 1980).

12. E. O. Wilson, "Human Decency is Animal," *New York Times Magazine*, 12 Oct. 1975.

13. Paul Weiss, "The Living System: Determinism Stratified," in *Beyond Reductionism: The Altbach Symposium*, ed. A. Koestler and J. R. Smithies (Boston: 1969).

14. J. A. Paton and F. N. Nottebohm, "Neurons Generated in the Adult Brain are Recruited into Functional Circuits," *Science* 225(1984):1046–48.

15. Roger Lewin, "Nutrition and Brain Growth," in *Child Alive*, ed. R. Lewin (New York: Anchor Books, 1975).

16. Dale Purves and J. W. Lichtman, "Elimination of Synapses in the Developing Nervous System," *Science* 210(1980):153–57.

17. T. N. Wiesel, "Postnatal Development of the Visual Cortex and the Influence of Environment," *Nature* 299(1982):592.

18. G. Ravelli, Z. Stein, and M. Susser, "Obesity in Young Men after Famine Exposure *in Utero* and Early Infancy," *New England Journal of Medicine* 295(1976): 349–53. See also J. C. Somogyi and H. Haenel, eds., *Nutrition in Early Childhood and its Effects in Later Life* (Basel: S. Karger, 1982).

19. W. S. Condon and L. W. Sander, "Neonate Movement is Synchronized with Adult Speech: Interactional Participation and Language Acquisition," *Science* 183(1974):99–101.

20. W. S. Condon and L. W. Sander, "Synchrony Demonstrated between Movements of the Neonate and Adult Speech," *Child Development* 45(1974):456–62.

21. C. M. Super, "Environmental Effects on Motor Development: The Case of 'African Infant Precocity,'" *Developmental and Medical Child Neurology* 18(1976): 561–67.

22. I. Frieze et al., *Women and Sex Roles: A Social Psychological Perspective* (New York: Norton, 1978).

23. S. A. Richardson, "The Relation of Severe Malnutrition in Infancy to the Intelligence of School Children with Differing Life Histories," *Pediatric Research* 10(1976):57–61.

24. G. S. Omenn and A. G. Motulsky, "Biochemical Genetics and the Evolution of Human Behavior," in *Genetics, Environment and Behavior: Implications for Educational Policy*, ed. L. Ehrmann, G. S. Omenn, and E. Caspari (New York: Academic, 1972), 131.

25. A. Ehrhardt and H. Meyer-Bahlburg, "Effects of Prenatal Sex Hormones on Gender-Related Behavior," *Science* 211(1981):1312–18.

26. R. Rubin, J. Reinsich, and R. Haskett, "Postnatal Gonadal Steroid Effects on Human Behavior," *Science* 211(1981):1318–24.

27. U. Muller and E. Urban, "Reaggregation of Rat Gonadal Cells *in vitro:* Experiments on the Function of H-Y Antigen," *Cytogenetics and Cellular Genetics* 31(1981):104–7; and U. Muller et al., "Ovarian Cells Participate in the Formation of Tubular Structures in Mouse/Rat Heterosexual Gonadal Co-Cultures," *Differentiation* 22(1982):136–38.

28. J. D. Wilson et al., "Recent Studies on the Endocrine Control of Male Phenotypic Development," in *Sexual Differentiation: Basic and Clinical Aspects*, ed. M. Serio et al. (New York: Raven, 1984).

29. B. H. Shapiro et al., "Neonatal Progesterone and Feminine Sexual Development," *Nature* 264(1976):795–96.

30. Wilson, George, and Griffin, "The Hormonal Control of Sexual Development," 1283.

31. Ibid.

32. Simone de Beauvoir, *The Second Sex* (New York: Knopf, 1952).

33. The following diagram shows the relationship of cortisone to testosterone/cholesterol and estrogen.

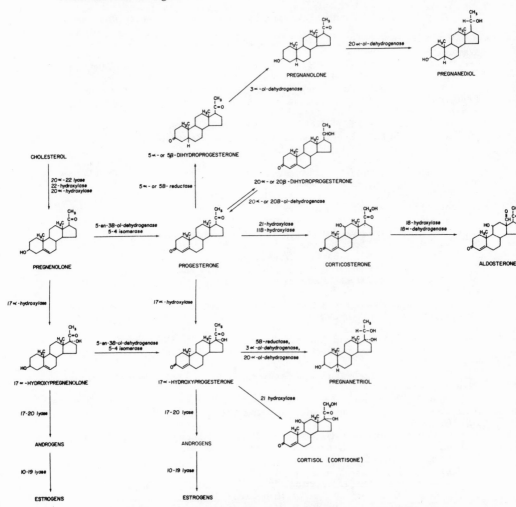

Biosynthesis of Steroid Hormones: Demonstrating the Chemical Relationship between Cholesterol, Progesterone, Cortisone, and the Estrogens and Androgens

NOTE: Harvey Leder, "Essentials of Steroid Structure, Nomenclature Reactions, Biosynthesis and Measurements," in *Neuroendocrinology of Reproduction*, ed. Norman T. Adler (New York: Plenum, 1981), 54–55. Reprinted by permission.

34. John Money, *Love and Lovesickness* (Baltimore: Johns Hopkins University Press, 1970), 5.

35. David Gelman et al., "Just How the Sexes Differ," *Newsweek*, 18 May 1981.

36. John Money and Anke A. Ehrhardt, *Man and Woman, Boy and Girl: Differentiation and Dimorphism of Gender Identity from Conception to Maturity* (Baltimore: Johns Hopkins University Press, 1972).

37. Julianne Imperato-McGuinley et al., "Steroid 5-Alpha-Reductase Deficiency in Man: An Inherited Form of Male Pseudohermaphroditism," *Science* 186(1974): 1213–15. See also Julianne Imperato-McGuinley et al., "Androgens and the Evolution of Male-Gender Identity among Male Pseudohermaphrodites with a 5-Alpha-Reductase Deficiency," *New England Journal of Medicine* 300 (1979):1236–37; and Letters in answer to Imperato-McGuinley et al., *New England Journal of Medicine* 301(1979): 839–40.

38. Imperato-McGuinley et al., "Androgens and the Evolution of Male-Gender Identity," 1235–36.

Chapter 4

1. Carroll Smith-Rosenberg and Charles Rosenberg, "The Female Animal: Medical and Biological Views of Woman and Her Role in 19th Century America," *Journal of American History* 60(1973):336.

2. Edgar Berman, Letter to the Editor, *New York Times*, 26 July 1970.

3. Steven Goldberg, *The Inevitability of Patriarchy* (New York: William Morrow, 1973), 93.

4. Herbert Wray, "Premenstrual Changes," *Science News* 122(1982):380–81.

5. Ilza Veith, *Hysteria: The History of a Disease* (Chicago: University of Chicago Press, 1965).

6. Pliny the Elder, quoted in M. E. Ashley-Montagu, "Physiology and Origins of the Menstrual Prohibitions," *Quarterly Review of Biology* 15(1940):211.

7. Smith-Rosenberg and Rosenberg, "The Female Animal"; Henry Maudsley, "Sex in Mind and in Education," *Popular Science Monthly* 5(1874):200; and Joan Burstyn, "Education and Sex: The Medical Case Against Higher Education for Women in England 1870–1900," *Proceeds of the American Philosophical Society* 117(1973):7989.

8. Carroll Smith-Rosenberg, "The Hysterical Woman: Sex Roles and Role Conflict in 19th Century America," *Social Research* 39(1972):652–78.

9. M. A. Hardaker, "Science and the Woman Question," *Popular Science Monthly* 20(1881):583.

10. Nina Morais, "A Reply to Ms. Hardaker on: The Woman Question," *Popular Science Monthly* 21(1882):74–75.

11. Maudsley, "Sex in Mind and in Education," 211.

12. Katharina Dalton, *Once a Month* (Claremont, Calif.: Hunter House, 1983), 78.

13. Ibid.; Katharina Dalton, *The Premenstrual Syndrome* (London: William Heinemann Medical Books, 1972).

14. Andrea Eagan, "The Selling of Premenstrual Syndrome," *Ms.* Oct. 1983, 26–31.

15. Robert L. Reid and S. S. Yen, "Premenstrual Syndrome," *American Journal of Obstetrics and Gynecology* 139(1981):86.

16. J. Abplanalp, R. F. Haskett, and R. M. Rose, "The Premenstrual Syndrome," *Advances in Psychoneuroendocrinology* 3(1980):327–47.

17. Dalton, *Once a Month*.

18. Boston Women's Health Collective, *Our Bodies, Ourselves* (New York: Simon and Schuster, 1979).

19. Reid and Yen, "Premenstrual Syndrome," 86.

20. John O'Connor, M. Shelley Edward, and Lenore O. Stern, "Behavioral Rhythms Related to the Menstrual Cycle," in *Biorhythms and Human Reproduction*, ed. M. Fern et al. (New York: Wiley, 1974), 312.

21. Frank A. Beach, Preface to chapter 10, in *Human Sexuality in Four Perspectives* (Baltimore: Johns Hopkins University Press, 1977), 271.

22. Rosenthal, "Insights into the Premenstrual Syndromes."

23. Herbert Schaumberg et al., "Sensory Neuropathy from Pyridoxine Abuse," *New England Journal of Medicine* 309(1983):446–48.

24. Judith Abplanalp, "Premenstrual Syndrome: A Selective Review," *Women and Health* 8(1983):110.

25. Dalton, *Once a Month*, 12.

26. Abplanalp, Haskett, and Rose, "The Premenstrual Syndrome"; and Abplanalp, "Premenstrual Syndrome: A Selective Review."

27. Abplanalp, "Premenstrual Syndrome: A Selective Review."

28. Reid and Yen, "Premenstrual Syndrome," 97.

29. G. A. Sampson, "An Appraisal of the Role of Progesterone in the Therapy of Premenstrual Syndrome," in *The Premenstrual Syndrome*, ed. P. A. vanKeep and W. H. Utian (Lancaster, England: MTP Press Ltd. International Medical Publishers, 1981), 51–69; and Sampson, "Premenstrual Syndrome: A Double-Blind Controlled Trial of Progesterone and Placebo," *British Journal of Psychiatry* 135(1979):209–15.

30. T. Benedek and B. B. Rubenstein, "The Correlations between Ovarian Activity and Psychodynamic Processes: I. The Ovulative Phase," *Psychosomatic Medicine* 1(1939):245–70; and T. Benedek and B. B. Rubenstein, "The Correlations between Ovarian Activity and Psychodynamic Processes: II. The Menstrual Phase, *Psychosomatic Medicine* 1(1939):461–85.

31. Janet R. Swanby, "A Longitudinal Study of Daily Mood Self Reports and their Relationship to the Menstrual Cycle," in *The Menstrual Cycle*, vol. 2, ed. P. Komnenich et al., (New York: Springer, 1981), 94.

32. Mary Putnam Jacobi, *The Question of Rest for Women During Menstruation* (New York: G. P. Putnam's Sons, 1877).

33. Mary Brown Parlee, "The Premenstrual Syndrome," *Psychological Bulletin* 80(1973):454–65.

34. Rudolf Moos, "Typology of Menstrual Cycle Symptoms," *American Journal of Obstetrics and Gynecology* 103(1969):390–402.

35. J. Delaney, M. J. Lupton, and E. Doth, *The Curse: A Cultural History of Menstruation* (New York: Dutton, 1976).

36. Randi Koeske, "Premenstrual Emotionality: Is Biology Destiny?," *Women and Health*, 1(1976):12.

37. Loudell Snow and Shirley M. Johnson, "Myths about Menstruation: Victims of Our Own Folklore," *International Journal of Women's Studies* 1(1978):70.

38. Ibid.

39. This study was reviewed by Sharon Golub, ed., "Lifting the Curse of Menstruation," *Women and Health* 8 (1983).

40. J. Brooks, D. N. Ruble, and A. E. Clark, "College Women's Attitudes and Expectations Concerning Menstrual-related Changes," *Psychosomatic Medicine* 39 (1977):288–98; and D. Ruble, "Premenstrual Symptoms: A Reinterpretation," *Science* 197(1977):291–92.

41. H. Persky, "Reproductive Hormones, Moods and the Menstrual Cycle," in *Sex Differences in Behavior*, ed. R. C. Friedman, R. M. Richart, and R. L. Vande Wiele (Huntington, N.Y.: Krieger, 1974), 455–66; W. P. Collins and J. R. Newton, "The Ovarian Cycle," in *Biochemistry of Women: Clinical Concepts*, ed. A. S. Curry and J. V. Hewitt (Boca Raton, Fla.: CRC Press, 1974).

42. Randi Koeske, "Premenstrual Emotionality: Is Biology Destiny?"; "Lifting the Curse of Menstruation: Toward a Feminist Perspective on the Menstrual Cycle," *Women and Health* 8(1983):1–16; and "Theoretical Perspectives on Menstrual Cycle Research," in *The Menstrual Cycle*, vol. 1, ed. A. Dan, E. Graham, and C. P. Beecher (New York: Springer, 1980), 8–24.

43. Koeske, "Theoretical Perspectives on Menstrual Cycle Research."

44. Koeske, "Lifting the Curse of Menstruation," 6.

45. Ibid., 13.

46. Koeske, "Theoretical Perspectives on Menstrual Cycle Research."

47. Alice Rossi and P. Rossi, "Body Time and Social Time: Mood Patterns by

Menstrual Cycle Phase and Day of the Week," *Social Science Research* 6(1977):273–308.

48. Koeske, "Theoretical Perspectives on Menstrual Cycle Research."

49. Alice Dan, "Free-Associative versus Self Report Measures of Emotional Change over the Menstrual Cycle," in *The Menstrual Cycle*, vol. 1, ed. Dan, Graham, and Beecher, 119.

50. Sharon Golub, "Premenstrual Changes in Mood, Personality and Cognitive Function," in *The Menstrual Cycle*, vol. 1, ed. Dan, Graham, and Beecher, 244.

51. See, for example, the volumes referred to in notes 31 and 32.

52. Koeske, "Theoretical Perspectives on Menstrual Cycle Research," 24.

53. Robert A. Wilson and Thelma A. Wilson, "The Fate of the Nontreated Postmenopausal Woman: A Plea for the Maintenance of Adequate Estrogen from Puberty to the Grave," *Journal of the American Geriatric Society* 11(1963):352–56.

54. David Reuben, *Everything You Always Wanted to Know about Sex but Were Afraid to Ask*. (New York: McKay, 1969), 292.

55. Wulf H. Utian, *Menopause in Modern Perspectives* (New York: Appleton-Century-Crofts, 1980).

56. Wilson and Wilson, "The Fate of the Nontreated Postmenopausal Woman," 347.

57. Robert A. Wilson, *Feminine Forever* (New York: M. Evans, 1966).

58. Marilyn Grossman and Pauline Bart, "The Politics of Menopause," in *The Menstrual Cycle*, vol. 1, ed. Dan, Graham, and Beecher.

59. Kathleen MacPherson, "Menopause as Disease: The Social Construction of a Metaphor," *Advances in Nursing Science* 3(1981):95–113; A. Johnson, "The Risks of Sex Hormones as Drugs," *Women and Health* 2(1977):8–11.

60. D. Smith et al., "Association of Exogenous Estrogen and Endometrial Cancer," *New England Journal of Medicine* 293(1975):1164–67.

61. H. Judd et al., "Estrogen Replacement Therapy," *Obstetrics and Gynecology* 58(1981):267–75; M. Quigley, "Postmenopausal Hormone Replacement Therapy: Back to Estrogen Forever?" *Geriatric Medicine Today* 1(1982):78–85; and Thomas Skillman, "Estrogen Replacement: Its Risks and Benefits," *Consultant* (1982):115–27.

62. C. Smith-Rosenberg, "Puberty to Menopause: The Cycle of Femininity in 19th Century America," *Feminist Studies* 1(1973):65.

63. Helene Deutsch, *The Psychology of Women* (New York: Grune and Stratton, 1945), 458.

64. J. H. Osofsky and R. Seidenberg, "Is Female Menopausal Depression Inevitable?," *Obstetrics and Gynecology* 36(1970):611.

65. Wilson and Wilson, "The Fate of the Nontreated Postmenopausal Woman," 348.

66. P. A. vanKeep, R. B. Greenblatt, and M. Albeaux-Fernet, eds., *Consensus on Menopause Research* (Baltimore: University Park Press, 1976), 134.

67. Marcha Flint, "Male and Female Menopause: A Cultural Put-on," in *Changing Perspectives on Menopause*, ed. A. M. Voda, M. Dinnerstein, and S. O'Donnell (Austin: University of Texas Press, 1982).

68. Utian, *Menopause in Modern Perspectives*.

69. *Ibid.*

70. Madeleine Goodman, "Toward a Biology of Menopause," *Signs* 5(1980): 739–53.

71. Madeleine Goodman, C. J. Stewart, and F. Gilbert, "Patterns of Menopause: A Study of Certain Medical and Physiological Variables among Caucasian and Japanese Women Living in Hawaii," *Journal of Gerontology* 32(1977):297.

72. Karen Frey, "Middle-Aged Women's Experience and Perceptions of Menopause," *Women and Health* 6(1981):31.

73. Eve Kahana, A. Kiyak, and J. Liang, "Menopause in the Context of Other Life Events," in *The Menstrual Cycle*, vol. 1, ed. Dan, Graham, and Beecher, 167–78.

74. Wilson, *Feminine Forever*, 134.

75. A. Voda and M. Eliasson, "Menopause: The Closure of Menstrual Life," *Women and Health* 8(1983):137–56.

76. Wilson and Wilson, "The Fate of the Nontreated Postmenopausal Woman," 356.

77. Louis Avioli, "Postmenopausal Osteoporosis: Prevention vs. Cure," *Federation Proceedings* 40(1981):2418.

78. Voda and Eliasson, "Menopause: The Closure of Menstrual Life."

79. G. Winokur and R. Cadoret, "The Irrelevance of the Menopause to Depressive Disease," in *Topics in Psychoendocrinology,* ed. E. J. Sachar (New York: Grune and Stratton, 1975).

80. Rosalind Barnett and Grace Baruch, "Women in the Middle Years: A Critique of Research and Theory," *Psychology of Women Quarterly* 3(1978):187–97.

81. Boston Women's Health Collective, *Our Bodies, Ourselves.*

82. D. Levinson et al., "Periods in the Adult Development of Men: Ages 18–45," *The Counseling Psychologist* 6(1976):21–25.

83. Barnett and Baruch, "Women in the Middle Years," 189.

84. Ibid.; Goodman, "Toward a Biology of Menopause"; and Voda, Dinnerstein, and O'Donnell, eds.,*Changing Perspectives on Menopause.*

Chapter 5

1. George Gilder, "The Case Against Women in Combat," *New York Times Magazine,* 28 Jan. 1979, p. 44.

2. Alexander Walker, *Woman Physiologically Considered* (New York: J. and H. G. Langley, 1850).

3. Steven Goldberg, *The Inevitability of Patriarchy* (New York: Morrow, 1973), 233–34.

4. Corinne Hutt, *Males and Females* (Harmondsworth, Middlesex: Penguin, 1972), 117.

5. David Barash, *The Whisperings Within* (New York: Harper and Row, 1979).

6. Ibid., 170, 171, 189, 192, 193.

7. Buss, A. H. "Aggression Pays" in *The Control of Aggression and Violence: Cognitive and Physiological Factors,* ed. J. Singer (New York: Academic Press, 1971), 7–18.

8. Arthur Kling, "Testosterone and Aggressive Behavior in Men and Non-Human Primates," in *Hormonal Correlates of Behavior,* ed. B. Eleftheriou and R. Spott (New York: Plenum, 1975), 305–23.

9. Tony Whitehead, "Sex Hormone Treatment of Prisoners," in *Multidisciplinary Approaches to Aggression Research,"* ed. P. Brain and D. Benton (Amsterdam: Elsevier, 1981), 503–11.

10. Kling, "Testosterone and Aggressive Behavior."

11. R. M. Rose, "Testosterone, Aggression and Homosexuality: A Review of the Literature and Implications for Future Research," in *Topics in Psychoendocrinology,* ed. E. J. Sachar (New York: Grune and Stratton, 1975), 95.

12. L. E. Kreuz and R. M. Rose, "Assessment of Aggressive Behavior and Plasma Testosterone in a Young Criminal Population," *Psychosomatic Medicine* 34(1972): 321.

13. Hutt, *Males and Females.*

14. W. Durham, "Resource Competition and Human Aggression, I: A Review of Primitive War," *Quarterly Review of Biology* 51(1976):385–415.

15. Francis Purifoy and Lambert Koopmans, "Androstenedione, Testosterone, and Free Testosterone Concentration in Women of Various Occupations," *Social Biology* 26(1980):179–88.

16. K. Matsumoto et al., "Plasma Testosterone Levels Following Surgical Stress in Male Patients," *Acta Endocrinologica* 65(1970):11.

17. L. E. Kreuz, R. M. Rose, and J. R. Jennings, "Suppression of Plasma Testosterone Levels and Psychological Stress," *Archives of General Psychiatry* 26(1972): 479–82; and Robert M. Rose, "Androgen Excretion in Stress," in *The Psychology and Physiology of Stress*, ed. Peter G. Bourne (New York: Academic, 1969).

18. Purifoy and Koopmans, "Androstenedione, Testosterone, and Free Testosterone Concentration," 185–86.

19. John Mason, "Organization of Endocrine Mechanisms," *Psychosomatic Medicine* 30(1968):796.

20. P. A. Jacobs et al., "Aggressive Behavior, Mental Abnormality and the XYY Male," *Nature* 208(1965):1351–52; and H. A. Witkin et al., "Criminality in XYY and XXY Men," *Science* 1983(1976):547–55.

21. Heino F. L. Meyer-Bahlburg, "Aggression, Androgens, and the XYY Syndrome," in *Sex Differences in Behavior*, ed. R. C. Friedman, R. M. Richart, and R. L. Vande Wiele (Huntington, N.Y.: Krieger, 1978), 447.

22. David Gelman, "Just How the Sexes Differ," *Newsweek*, 18 May 1981, 74.

23. J. R. Reinisch, "Fetal Hormones, the Brain and Human Sex Differences: A Heuristic, Integrative Review of the Recent Literature," in *Archives of Sexual Behavior* 3(1974):51.

24. S. K. Ratzan and V. V. Weldon, "Exposure to Endogenous and Exogenous Sex Hormones During Pregnancy," in *Influence of Maternal Hormones on the Fetus and Newborn. Pediatric and Adolescent Endocrinology*, vol. 5 (Basel: Karger, 1979), 186.

25. John Money and Anke Ehrhardt, *Man and Woman, Boy and Girl* (Baltimore: Johns Hopkins University Press, 1972).

26. P. Farnes and R. Hubbard, "Clitoridectomy in the U.S.," *Science for the People* 13(1981):2.

27. Money and Ehrhardt, *Man and Woman, Boy and Girl*, 99, 103.

28. Anke Ehrhardt and S. W. Baker, "Fetal Androgens, Human Central Nervous System Differentiation, and Behavior Sex Differences," in *Sex Differences in Behavior*, ed. R. Friedman, R. M. Richart, and R. L. Vande Wiele (Huntington, N.Y.: Krieger, 1978).

29. Ruth Bleier, *Science and Gender* (Elmsford, N.Y.: Pergamon, 1984).

30. Ehrhardt and Baker, "Fetal Androgens," 49.

31. Heino F. L. Meyer-Bahlberg and Anke Ehrhardt, "Prenatal Sex Hormones and Human Aggression: A Review," *Aggressive Behavior* 8(1982):59.

32. M. Richards, J. Bernal, and Y. Brackbill, "Early Behavioral Differences: Gender or Circumcision?," *Developmental Psychology* 9(1976):89–95.

33. Laura S. Sidorowicz and G. Sparks Lunney, "Baby X Revisited," *Sex Roles* 6(1980):67–73.

34. John Money, J. G. Hampson, and J. L. Hampson, "An Examination of Some Basic Sexual Concepts: The Evidence of Human Hermaphroditism," *Johns Hopkins Medical Journal* 97(1955):301–19.

35. June Reinisch, "Prenatal Exposure to Synthetic Progestins Increases Potential for Aggression in Humans," *Science* 211(1981):1171–73.

36. Anke A. Ehrhardt and Heino F. L. Meyer-Bahlburg, "Prenatal Sex Hormones and the Developing Brain," *Annual Review of Medicine* 30(1979):417–30.

37. Reinisch, 1173.

38. R. B. Heap, "Role of Hormones in Pregnancy," in *Hormones in Reproduction*, ed. C. R. Austin and R. V. Short (New York: Cambridge University Press, 1972).

39. Ellen Farber, B. Vaughn, and B. Egeland, "The Relationship of Prenatal Maternal Anxiety to Infant Behavior and Mother-Infant Interaction During the First Six Months of Life," *Early Human Development* 5(1981):267–77.

40. Clara Torda, "Early Responses to Stress and Subsequent Behavioral Potential," in *Multidisciplinary Approaches to Aggression Research*, ed. P. F. Brain and D. Benton (New York: Elsevier, 1981).

41. Anke Ehrhardt and Heino F. L. Meyer-Bahlburg, "Effects of Prenatal Sex Hormones on Gender-related Behavior," *Science* 211(1981):1312–17.

42. I. Yalom, R. Green, and N. Fisk, "Prenatal Exposure to Female Hormones—Effect on Psychosexual Development in Boys," *Archives of General Psychiatry* 28(1973):554–61.

43. Charles Phoenix, "Prenatal Testosterone in the Non-Human Primate and its Consequences for Behavior," in *Sex Differences in Behavior*, ed. Richard Friedman, R. Richart, and R. L. Vande Wiele (Huntington, N.Y.: Krieger, 1978).

44. T. Rowell, *Social Behavior of Monkeys* (Harmondsworth, Middlesex: Penguin, 1972).

45. R. W. Goy and J. A. Robinson, "Prenatal Exposure of Rhesus Monkeys to Patent Androgens," in *Banbury Report*, vol. 11 (Cold Spring Harbor, N.Y.: Cold Spring Harbor Laboratory, 1982), 355–78.

46. G. Mitchell and E. M. Brandt, "Behavioral Differences Relate to Experience of Mother and Sex of Infant in the Rhesus Monkey," *Developmental Psychology* 3(1970):149.

47. I. Bernstein, "Dominance: The Baby and the Bathwater," *Behavioral and Brain Sciences* 4(1981):419–57.

48. Robert Ardrey, *African Genesis* (New York: Dell, 1967), 91.

49. G. B. Kolata, "Primate Behavior: Sex and the Dominant Male," *Science* 191(1976):55–56.

50. Ibid.; T. Rowell, *Social Behavior of Monkeys;* T. Rowell, "The Concept of Social Dominance," *Behavior Biology* 11(1974):131–54; and Bernstein, "Dominance: The Baby and the Bathwater."

51. Rowell, "The Concept of Social Dominance."

52. Kolata, "Primate Behavior"; Rowell, "The Concept of Social Dominance"; Linda M. Fedigan, *Primate Paradigms* (Montreal: Eden, 1982).

53. A. F. Dixson, "Androgens and Aggressive Behavior in Primates: A Review," *Aggressive Behavior* 6(1980):37–67.

54. Rowell, "The Concept of Social Dominance."

55. J. L. Popp, "Male Baboons and Evolutionary Principles" (Ph.D. diss., Harvard University, 1978).

56. G. G. Eaton and J. A. Resko, "Plasma Testosterone and Male Dominance in a Japanese Macaque Troop Compared with Repeated Measurements of Testosterone in Laboratory Males," *Hormones and Behavior* 5(1974):257.

57. T. P. Gordon, R. M. Rose, and I. S. Bernstein, "Seasonal Rhythm in Plasma Testosterone Levels in the Rhesus Monkey: A Three Year Study," *Hormones and Behavior* 7(1976):229–43.

58. R. M. Rose, T. P. Gordon, and I. S. Bernstein, "Plasma Testosterone Levels in Male Rhesus: Influence of Sexual and Social Stimuli," *Science* 178(1972):643–45.

59. R. Rose, "Androgen Excretion in Stress," in *The Psychology and Physiology of Stress*, ed. P. Bourne (New York: Academic, 1969), 117–48.

60. Cheryl Harding, "Social Modulation of Circulating Hormone Levels in the Male," *American Zoologist* 21(1981):223–31.

61. Kenneth Moyer, "Kinds of Aggression and their Physiological Basis," *Communications in Behavioral Biology* 2(1968):65–87.

62. Eleanor Maccoby and Carol N. Jacklin, *The Psychology of Sex Differences* (Stanford, Calif.: Stanford University Press, 1974).

63. A. Frodi, J. Macaulay, and P. R. Thome, "Are Women Always Less Aggressive than Men?," *Psychological Bulletin* 84(1977):634–60.

64. Ibid., 274.

65. Todd Tieger, "On the Biological Basis of Sex Differences in Aggression," *Child Development* 51(1980):943–63.

66. J. Condry and S. Condry, "Sex Differences: A Study in the Eye of the Beholder," *Child Development* 47(1976):817.

67. Tieger, "Biological Basis of Sex Differences."

68. Carol Ember, "Feminine Task Assignment and the Social Behavior of Boys," *Ethos* 1(1973):424–39.

69. Tieger, "Biological Basis of Sex Differences," 959.

70. R. A. Dart, "The Predatory Transition from Ape to Man," *International Anthropological and Linguistic Review* 1(1953):209.

238

71. S. Washburn and C. Lancaster, "The Evolution of Hunting," in *Man the Hunter*, ed. R. B. Lee and I. De Vore (Chicago: Aldine, 1968).

72. Bleier, *Science and Gender*; N. Tanner, *On Becoming Human* (New York: Cambridge University Press, 1981); L. Leibowitz, "Origins of the Sexual Division of Labor," in *Woman's Nature*, ed. M. Lowe and R. Hubbard (Elmsford, N.Y.: Pergamon, 1983); and Ruth Bleier, *Sex and Gender* (Elmsford, N.Y.: Pergamon, 1984).

Chapter 6

1. David Barash, *The Whisperings Within: Evolution and the Origin of Human Nature* (New York: Harper and Row, 1979), 54.

2. Randy Thornhill, "Rape in Panorpa Scorpionflies and a General Rape Hypothesis," *Animal Behavior* 28(1980):57.

3. Richard D. Alexander and K. M. Noonan, "Concealment of Ovulation, Parental Care and Human Social Evolution," in *Evolutionary Biology and Human Social Behavior*, ed. N. Chagnon and William Irons (North Scituate, Mass.: Duxbury, 1979), 449.

4. Donald Symons, *The Evolution of Human Sexuality* (New York: Oxford University Press, 1979), 284–85.

5. Edward O. Wilson, *Sociobiology: The New Synthesis* (Cambridge, Mass.: Harvard University Press, 1975), 4.

6. Robert Trivers, "Sociobiology and Politics," in *Sociobiology and Human Politics*, ed. Elliott White (Lexington, Mass.: Heath, 1981).

7. Richard D. Alexander, *Darwinism and Human Affairs* (Seattle: University of Washington Press, 1980).

8. Robin Fox, "Kinship Categories as Natural Categories," in *Evolutionary Biology and Human Social Behavior*, ed. Chagnon and Irons.

9. M. Dickerman, "Female Infanticide, Reproductive Strategies, and Social Stratification: A Preliminary Model," in *Evolutionary Biology and Human Social Behavior*, ed. Chagnon and Irons.

10. Lionel Tiger, "The Possible Biological Origins of Sexual Discrimination," *Impact of Science on Society* 20(1970):29–44.

11. David Barash, "Predictive Sociobiology: Mate Selection in Damselfishes and Brood Defense in White-Crowned Sparrows," in *Sociobiology: Beyond Nature/Nurture?*, ed. G. W. Barlow and J. Silverberg (Boulder, Colo.: Westview, 1980), 212.

12. Thornhill, "Rape in Scorpionflies."

13. Randy Thornhill and N. Thornhill, "Human Rape: An Evolutionary Analysis," *Ethology and Sociobiology* 4(1983):137.

14. David Barash, "Sociobiology of Rape in Mallards: Responses of the Mated Male," *Science* 197(1977):788.

15. E. O. Wilson, "Academic Vigilantism and the Political Significance of Sociobiology," *BioScience* 26(1976):187.

16. Barash, *The Whisperings Within*, 54.

17. Randy Thornhill, "Sexually Selected Predatory and Mating Behavior of the Hangingfly *Bittacua Stignaterus*," *Annals of the Entomological Society of America* 71(1978):597–601.

18. Randy Thornhill, "Rape in Scorpionflies and a General Rape Hypothesis," *Animal Behavior* 28(1980):52.

19. L. G. Abele and S. Gilchrist, "Homosexual Rape and Sexual Selection in Acanthocephalan Worms," *Science* 197(1977):81–83.

20. Thornhill, "Rape in Scorpionflies," 55.

21. Ibid., 55–56.

22. Ibid., 56.

23. Thornhill and Thornhill, "Human Rape," 141.

24. Ibid.

25. W. M. Shields and L. M. Shields, "Forcible Rape: An Evolutionary Perspective," *Ethology and Sociobiology* 4(1983):115–36.

26. Charles Darwin, *The Origin of Species by Means of Natural Selection* (1859; reprint, New York: Mentor, 1958), 87–88.

27. Julian Huxley, *Evolution: The Modern Synthesis* (New York: Harper Bros., 1943).

28. G. Ledyard Stebbins, *Darwin to DNA, Molecules to Humanity* (San Francisco: Freeman, 1982).

29. Ibid.

30. Stephen Jay Gould and R. C. Lewontin, "The Spandrels of San Marco and the Panglossian Paradigm: A Critique of the Adaptationist Programme, *Proceedings of the Royal Society of London B* 205(1979):581–98.

31. Richard C. Lewontin, *The Genetic Basis of Evolutionary Change* (New York: Columbia University Press, 1974).

32. Richard Lewontin, "Adaptation," *Scientific American* 239(1978):212–30.

33. Sewall Wright, "The Roles of Mutation, Inbreeding, Crossbreeding and Selection in Evolution," in *Proceedings of the Sixth International Congress of Genetics* (ed. I. F. Jones), 1(1932):356–66.

34. Gould and Lewontin, "The Spandrels of San Marco."

35. Ibid.

36. Huxley, *Evolution: The Modern Synthesis*, 18.

37. H. Kettlewell, "Darwin's Missing Evidence," *Scientific American* 200(1959): 48–53.

38. H. Kettlewell, "A Resume of Investigations on the Evolution of Melanism in the *Lepidoptera*," *Proceedings of the Royal Society of London B* 145(1956):297–303.

39. Dian Fossey, *Gorillas in the Mist* (Boston: Houghton Mifflin, 1983).

40. Barash, "Predictive Sociobiology," 211.

41. Wilson, *Sociobiology: The New Synthesis.*

42. Barash, "Predictive Sociobiology," 212.

43. Barash, "The Whisperings Within," 120–21.

44. Janet Sayers, *Biological Politics* (London: Tavistock, 1982); Antoinette Brown Blackwell, *The Sexes Throughout Nature* (New York: Putnam, 1875; Westport, Conn.: Hyperion, 1976); Elizabeth Gamble, *Evolution and Woman: An Inquiry into the Dogma of Her Inferiority to Man* (London: Putnam, 1893; Westport, Conn.: Hyperion, 1976); and Frances Emily White, "Women's Place in Nature," *Popular Science Monthly* 6(1875):292–301.

45. Charles Darwin, *The Descent of Man, and Selection in Relation to Sex* (1871; reprint, Princeton, N.J.: Princeton University Press, 1981).

46. Alfred Russel Wallace, *Darwinism: An Exposition of the Theory of Natural Selection* (London: AMS, 1889); and Alfred Russel Wallace, *Tropical Nature and Other Essays* (London: AMS, 1878).

47. Jane Lancaster, *Primate Behavior and the Emergence of Human Culture* (New York: Holt, Rinehart and Winston, 1975).

48. Huxley, *Evolution: The Modern Synthesis*, 525.

49. Ernst Mayr, "Sexual Selection and Natural Selection," in *Sexual Selection and the Descent of Man*, ed. Bernard Campbell (Chicago: Aldine, 1972), 97–98, 320.

50. Wilson, *Sociobiology: The New Synthesis*, 320.

51. David Barash, *Sociobiology and Behavior* (New York: Elsevier, 1977), 147.

52. Patrick Geddes and J. Arthur Thompson, *The Evolution of Sex* (New York: Humboldt, 1890).

53. R. Selander, "Sexual Selection and Dimorphism in Birds," in *Sexual Selection and the Descent of Man*, ed. Campbell.

54. Robert Trivers, "Parental Investment and Sexual Selection," in *Sexual Selection and the Descent of Man*, ed. Campbell, 139, 173.

55. Ibid., 145.

56. Wilson, *Sociobiology: The New Synthesis*, 327.

57. Ibid.

58. Richard C. Lewontin, "Sociobiology as an Adaptationist Program," *Behavioral Science* 24(1979):9.

59. Ibid., 10.

60. Thornhill, "Rape in Scorpionflies," 53.

61. Randy Thornhill, "*Panorpa* Scorpionflies: Systems for Understanding Resource-Defense Polygyny and Alternative Male Reproductive Efforts," *Annual Reviews of Ecology and System Matters* 12(1981):355–86.

62. Barash, "Sociobiology of Rape in Mallards," 788, 789.

63. J. P. Hailman, "Rape Among Mallards," *Science* 201(1978):280–81; and F. McKinney, J. Barrett, and S. R. Derrickson, "Rape Among Mallards," *Science* 201(1978):281–82.

64. R. D. Titman and J. K. Lowther, "The Breeding Behavior of a Crowded Population of Mallards," *Canadian Journal of Zoology* 53(1975):1270–82.

65. T. Lebret, "The Pair Formation in the Annual Cycle of the Mallard, *Anas Platyrhynchos*," *Ardea* 49(1961):97–158.

66. Shields and Shields, "Forcible Rape: An Evolutionary Perspective," 119, 120.

67. Ibid., 122.

68. Ibid., 132.

69. Ibid.

70. "Judge Orders 'Castration Drug' for Upjohn Heir," *The Providence Journal*, 31 Jan. 1984, pp. x–1.

71. Phillida Bunkle, "Calling the Shots? The International Politics of Depo-Provera," in *Test Tube Women: What Future for Motherhood*, ed. Rita Arditti, Renate Duelli Klein, and Shelley Minden (London: Pandora, 1984), 165–87.

72. CBS-TV, *60 Minutes*, vol. 16, no. 18, 15 Jan. 1984. Transcript.

73. Wilson, *Sociobiology: The New Synthesis*, 548.

74. B. A. Hamburg, "The Psychobiology of Sex Differences," in *Sex Differences in Behavior*, ed. R. C. Friedman, R. M. Rienhart, and J. R. L. Van de Wiele (New York: Wiley, 1974).

75. Edward O. Wilson, *On Human Nature* (Cambridge, Mass.: Harvard University Press, 1978).

76. Donald Symons, "Précis of *The Evolution of Human Sexuality*," *The Behavioral and Brain Sciences* 3 (1980):176, 179.

77. Robert Trivers, "Foreword," in R. Dawkins, *The Selfish Gene* (New York: Oxford University Press, 1976).

78. Wilson, *On Human Nature*.

79. Trivers, "Sociobiology and Politics," 37.

80. R. Passingham, *The Human Primate* (San Francisco: Freeman, 1982).

81. Michelle Rosaldo, "The Use and Abuse of Anthropology," *Signs* 5(1980): 389–417.

82. Karen Sacks, *Sisters and Wives: The Past and Future of Sexual Inequality* (Westport, Conn.: Greenwood, 1979); C. MacCormack and M. Strathern, eds., *Nature, Culture and Gender* (New York: Cambridge University Press, 1980); S. Ortner and H. Whitehead, eds., *Sexual Meanings: The Cultural Construction of Gender and Sexuality* (New York: Cambridge University Press, 1981); and Eleanor Leacock, *Myths of Male Dominance* (New York: Monthly Review, 1981).

83. Stebbins, *Darwin to DNA, Molecules to Humanity*, 399.

84. Barash, *The Whisperings Within*, 235.

85. Trivers, "Foreword," vi.

86. Symons, *The Evolution of Human Sexuality*, 284, 285.

87. Wilson, *On Human Nature*, 132, 133.

88. Ibid., 135.

89. John MacKinnon, *The Ape Within Us* (New York: Holt, Rinehart and Winston, 1978), 264–65.

90. In the October 1977 issue of *Anthropology Newsletter*, Wilson, DeVore, and Trivers wrote that they had nothing to do with making the film aside from providing interviews. They said, "We deplore the vulgar misrepresentation of the field by those who use this discredited film to imply that it represents an accurate statement of our ideas." This quote is not from the original source but is quoted in

Ted Judd, "Naturizing What We Do: A Review of the Film *Sociobiology: Doing What Comes Naturally*," *Science for the People*, Jan./Feb. 1978, 19.

91. M. T. McGuire, "Sociobiology: Its Potential Contributions to Psychiatry," *Perspectives in Biology and Medicine* 25(1979):50–69.

92. Trivers, "Sociobiology and Politics," 37.

93. Ibid.

94. Wilson, *On Human Nature*, 3.

95. Ibid.

96. C. J. Lumsden and E. O. Wilson, "Genes, Mind and Ideology," *The Sciences*, Nov. 1981, 8.

Chapter 7

1. Robin Fox, quoted in E. O. Wilson, *Sociobiology: The New Synthesis* (Cambridge, Mass.: Harvard University Press, 1975), 569.

2. Stephen Jay Gould, *The Mismeasure of Man* (New York: Norton, 1981); Marian Lowe and Ruth Hubbard, *Woman's Nature: Rationalizations of Inequality* (New York: Pergamon, 1983); and Ruth Bleier, *Gender and Science* (New York: Pergamon, 1977).

3. Stephen Goldberg, *The Inevitability of Patriarchy* (New York: Morrow, 1974).

4. Edward O. Wilson, *On Human Nature* (Cambridge, Mass.: Harvard University Press, 1978), 125.

5. Stephen Jay Gould, *New York Times Book Review*, 12 Aug. 1984, p. 7.

6. See, for example, the articles by Naomi Weisstein and Evelyn Fox Keller in *Working It Out*, ed. S. Ruddick and P. Daniels (New York: Pantheon, 1977).

7. Randi D. Koeske, "Toward a Biosocial Paradigm for Menopause Research: Lessons and Contributions from the Behavioral Sciences," in *Changing Perspectives on Menopause*, ed. A. Voda, M. Dinnerstein, and S. O'Donnell (Austin: University of Texas Press, 1982).

8. Ibid., 11–12.

9. Ibid., 16.

10. Ibid., 1.

11. A. Walker, *Woman Physiologically Considered* (New York: J. and H. G. Langley, 1843), 8.

12. P. Weiss, *Sport: A Philosophical Inquiry* (Carbondale, Ill.: Southern Illinois University Press, 1969), 215.

13. Ida Pruitt, *A Daughter of Han* (New Haven: Yale University Press, 1945), 22.

14. S. B. Stromme, D. Meen, and A. Arkvaag, "Effects of an Androgenic-Anabolic Steroid on Strength Development and Plasma Testosterone Levels in Normal Males," *Medicine and Science in Sports* 6(1974):203–8.

15. J. R. Sutton et al., "Hormonal Changes During Exercise," *Lancet* 2(1968): 1304; and J. Roth et al., "Secretion of Human Growth Hormone: Physiologic and Experimental Modification," *Metabolism* 12(1963):577–79.

16. G. M. Andrew et al., "Heart and Lung Functions in Swimmers and Non-Athletes During Growth," *Journal of Applied Physiology* 32(1972):245–61.

17. J. H. Wilmore, "They Told Me You Couldn't Compete with Men and You, Like a Fool, Believed Them. Here's Hope," *Women Sports* 1(1974):40–43, 83.

18. J. Wilmore, C. H. Brown, and J. A. Davis, "Body Physique and Composition of the Female Distance Runner," *Annals of the New York Academy of Sciences* 301(1977):764–76.

19. H. J. Wilmore and C. H. Brown, "Physiological Profiles of Women Distance Runners," *Medicine and Science in Sports* 6(1974):173–81.

20. Stromme, Meen, and Arkvaag, "Effects of an Androgenic-Anabolic Steroid."

21. Margaret Mead, *Male and Female* (New York: Morrow, 1949), 175.

22. J. Hudson, "Physical Parameters Used for Female Exclusion from Law Enforcement and Athletics," in *Women and Sport: From Myth to Reality*, ed. Carole A. Oglesby (Philadelphia: Lea and Febiger, 1978).

23. S. L. Twin, *Out of the Bleachers* (Westbury, N.Y.: Feminist Press, 1979); and Nina Kusick, "The History of Women's Participation in the Marathon," *Annals of the New York Academy of Science* 301(1977):862–76.

24. Hudson, "Physical Parameters Used for Female Exclusion."

25. Ken Dyer, "Female Athletes are Catching Up," *New Scientist*, 22 Sept. 1977, 722–23.

26. Lynda Birke, "Is Homosexuality Hormonally Determined?," *Journal of Homosexuality* 6(1981):35–50; Michael Ruse, "Are there Gay Genes? Sociobiology and Homosexuality," *Journal of Homosexuality* 6(1981):5–34; J. Hopkins, M. Marcus, and S. B. Campbell, "Post-Partum Depression: A Critical Review," *Psychological Bulletin* 95(1984):498–515; L. Hoffman, R. Grendelman, and H. Schiffman, eds., *Parenting: Its Causes and Consequences* (Hillsdale, N.J.: Erlbaum, 1982); and Janice Raymond, *The Transsexual Empire* (Boston: Beacon, 1979).

27. Sheila Tobias, "Equity, Productivity and the So-called 'Crisis in Science and Mathematics Education,'" *Association for Women in Science Newsletter* 13(1984): 11.

Appendix

1. Richard Lewontin, *The Genetic Basis of Evolutionary Change* (New York: Columbia University Press, 1974), 236.

2. Ibid.

3. R. Lewin, "Red Deer Data Illuminate Sexual Selection," *Science* 218(1982): 1207–8.

GLOSSARY

Adaptation (in evolution): Some inherited part of an individual's phenotype that improves its chances of survival and reproduction in the existing environment. *In general biological use:* Some part of an individual's phenotype (not necessarily inherited) that improves its chances of survival and reproduction in the existing environment.

Adrenal glands: Endocrine glands located near the kidneys that are responsible for the synthesis of a wide variety of hormones. Adrenal glands are especially involved in the synthesis of steroid hormones, including both androgens and estrogens.

Adrenogenital syndrome (AGS): An inherited defect in cortisone biosynthesis which results in the overproduction of certain androgens. In XX females it can result in the masculinization of female genitalia.

Allele: One of two or more forms that can exist at a single gene locus, and which are distinguished by their different effects on phenotype.

Amino acid: The basic building block of proteins. There are more than twenty different amino acids, each constructed on a similar chemical skeleton.

Androgen: A hormone that promotes the development and maintenance of male secondary sex characteristics.

Androgen insensitivity syndrome (AIS): A genetically inherited metabolic error which prevents the cells of the body from binding testosterone. XY individuals with this error develop as females because their bodies cannot respond to the masculinizing effects of testosterone. (Formerly called *testicular feminization*.)

Androstenedione: An androgen that may be converted to either testosterone or estrone by a specific enzyme. It may be synthesized in either the adrenal glands or the testes.

Autosome: Any chromosome that is not a sex chromosome.

Cerebellum: The lower part of the brain lying beneath the cerebrum. A major function of the cerebellum is to regulate muscular coordination, especially of the various muscles involved in voluntary responses (such as typing these glossary words).

Cerebrum: The largest portion of the brain, occupying the whole upper part of the cranium and consisting of left and right hemispheres. The *cerebral cortex* is involved in the control of motor and cognitive functions.

Chromosome: A linear end-to-end arrangement of genes and other DNA in association with proteins and RNA.

Climacteric: The phase in the aging process of women that marks the transition from the reproductive stage to the nonreproductive stage of life.

Corpus callosum: A large mass of nerve fibers that connect the left and right hemispheres of the cerebral cortex.

Corpus luteum: A specialized glandular structure within the ovary, responsible for synthesizing progesterone. In the human menstrual cycle the corpus luteum is active during the postovulatory phase. During pregnancy the corpus luteum remains active and grows to considerable size within the ovary.

Cortisone: A major steroid hormone synthesized by the adrenal glands. The hormone progesterone is a precursor in the pathway leading to cortisone synthesis. In patients suffering from the adrenogenital syndrome (AGS), cortisone is used as a treatment to regulate the activity of the adrenal glands and to control the overproduction of testosterone.

Crossing over: The exchange of corresponding chromosome parts between homologous chromosomes.

Dihydrotestosterone (DHT): An androgen synthesized from testosterone by the action of the enzyme 5-alpha-reductase. In XY individuals deficient in this enzyme, the external genitalia develop abnormally in the fetus and newborns may be mistaken for females. At puberty masculine characteristics develop because pubertal masculinization is influenced by testosterone. (Fetal masculinization is influenced by dihydrotestosterone.)

Dimorphism (sexual): The presence of two sexes of different phenotypes in a species.

DNA (Deoxyribonucleic acid): A double chain of linked bases and sugars; the fundamental substance of which genes are made.

Dominant allele: An allele that expresses its phenotypic effect even when heterozygous with a recessive allele; if *A* is dominant over *a*, then *AA* and *Aa* have the same phenotype.

Enzyme: A protein that functions as a catalyst. It aids in the carrying out of a chemical reaction without being changed itself by the reaction.

Epigamic: Any trait related to courtship and sex other than the essential organs and behavior of copulation.

Estradiol: An estrogen derived from testosterone.

Estriol: An estrogen derived from estrone.

Estrogens: Hormones found principally in the ovaries and the placenta which stimulate development of the accessory sex structures and secondary sex characteristics in the female.

Estrone: An estrogen that may be synthesized from the precursor androstenedione or from the precursor estradiol.

Estrus: The period of maximum sexual receptivity in female animals; usually coincides with the time of ovulation.

Fitness (individual or Darwinian): The relative probability of survival and reproduction for an individual of a particular genotype.

Follicle-stimulating hormone (FSH): A peptide hormone synthesized by cells of the pituitary gland. In females it is involved in stimulating the growth of ovarian egg follicles and releasing the ripened egg from the ovary. In males it is involved in the control of sperm growth and development.

Gene: The physical unit of heredity which one recognizes through its variant alleles; in most organisms it consists of a specific region of DNA.

Gene pool: The sum of all the variations found in the genes of individuals that compose a particular population.

Genome: The sum of all the different chromosomes found in the nuclei of the cells of a particular individual. Except for identical twins, no two individuals have identical genomes.

Genotype: The specific allele composition of an organism, either of the entire genome or more typically for a certain gene or set of genes.

Glial cells: Specialized brain cells involved with myelin synthesis and with other functions that are not clearly understood.

Hemoglobin: The protein molecule in red blood cells that carries oxygen throughout the body. *Hemoglobin A* is the normal form of the protein found in adults; *Hemoglobin S* is the abnormal hemoglobin protein found in people with sickle-cell anemia.

Hermaphrodite: A plant or animal having the reproductive organs of both sexes.

Heterozygote: An organism having gene pairs with different alleles in the two chromosome sets, usually one dominant and one recessive (*Aa*). (Adjective: **heterozygous.**)

Homologues (homologous chromosomes): Chromosomes that are identical in shape, size, and function and that pair with one another during meiosis.

Homozygote: An organism having a homozygous gene pair (identical alleles in both copies), e.g., AA or A'A'.

Hormones (peptide): Hormones composed of short sequences of amino acids. Follicle-stimulating hormones and luteinizing hormones are examples.

Hormones (steroid): Lipid hormones derived from the molecule cholesterol. Steroid hormones include the estrogens, the androgens, and hormones such as cortisone.

H-Y antigen: A protein that can induce formation of antibodies and that, in mammals, is associated with the presence of the Y chromosome. The H-Y antigen relates strongly with maleness in mammals, but its exact role in sex determination is unknown.

Hypothalamus: A specialized portion of the brain which produces a number of chemical releasing factors that stimulate the pituitary gland to secrete certain hormones.

Hypothesis testing: A form of statistical inference. In the study of sex differences a researcher hypothesizes the existence of a difference, then tests for it, calculates average scores for males and females, and uses a statistical test to compare the means and standard deviations found for the male and female groups. The statistical test determines whether any difference found is statistically significant (see *statistical significance*).

Indifferent gonad: A stage of embryonic gonadal development during which the gonad is identical in all embryos, regardless of sex. It is bipotential, i.e., it can develop into either a male or a female gonad depending upon the hormonal environment in which it is found.

Industrial melanism: A change in the gene pool of the moth *Biston betularia* in industrialized regions of Europe. The population shifted from predominantly lightly-colored moths in nonindustrial regions to darkly-colored ones in regions in which industrial pollution darkened the tree bark on which the moths sit.

Kin selection: The selection of genes favoring or disfavoring the survival and reproduction of relatives (other than offspring) who possess the same genes by common descent.

Lipid: A fat, or fat derivative. Chemists note that lipids are generally insoluble in water but soluble in oil or other fats.

Luteinizing hormone (LH): A peptide hormone synthesized by cells of the pituitary gland. In females it is involved in the release of the ripened egg from the ovary and the subsequent growth of the corpus luteum. In males—where it has been renamed **interstitial cell stimulating hormone**—it stimulates the interstitial cells of the testes to synthesize testosterone.

Meiosis: A special cell division occurring during the process of egg and sperm formation in which the number of chromosomes is cut in half. The chromosome number is doubled again at the time of fertilization.

Menopause: Indicator of the final menstrual period. It occurs during the climacteric.

Meta-analysis: A statistical method of analyzing differences between two groups. Instead of looking separately at the two groups to be compared (see *hypothesis testing*), the entire population (e.g., males and females) is looked at together. One then estimates the variability of the entire population and calculates how much of it is due to a particular variable (such as gender or course background).

MPA (medroxyprogesterone acetate): A synthetic progesterone used to lower testosterone production. A so-called "chemical castration" agent, used in some studies to feminize the embryo.

Müllerian inhibiting substance (MIS): A hormone normally produced in XY embryos by the fetal testis. MIS promotes the degeneration of the Müllerian ducts, embryonic structures in XX embryos that develop into the fallopian tubes and uterus.

Mutation: A process that produces a gene or a chromosome set differing from the norm; the gene or chromosome resulting from that process.

Myelin: A white, fatty substance that forms a sheath around some nerve cells and apparently serves as an electrical insulator. Cells with myelin sheaths form the white mattter of the cerebral cortex.

Natural selection: A process based on overproduction and genetic variability, which leads to the differential survival of the better-adapted organisms in a particular environment; a process that maintains or alters gene frequencies in a population.

Neuron: A complete nerve cell including the cell body, axon, and dendrites; it is a specialized conductor of electrical impulses.

19-NET (19-nor-17-alpha-ethyltestosterone): A synthetic progesterone that, when given in high doses to pregnant animals, can cause the masculinization of developing female genitalia.

Nucleic acid bases: Nitrogen-containing molecues that form key parts of nucleic acids (DNA and RNA). Weak chemical bonds between bases link the strands of a DNA double helix.

Nucleus: Intracellular structure bound by a membrane and containing the chromosomal DNA of the cell. During mitosis (cell division) the DNA condenses into individually visible chromosomes and is distributed equally to the two daughter cells.

Parasympathetic: See *sympathetic nervous system.*

Parental investment: Any behavior on the part of a parent toward its offspring that increases the chances of the offspring's survival at the cost of the parent's ability to raise other offspring.

Peptide: A combination of two or more amino acids joined by a particular kind of chemical connection called a "peptide linkage."

Phenotype: The form taken by some character or group of characters in a specific individual, or the detectable outward manifestation of a particular genotype.

Phenyl alanine: An essential amino acid, i.e., one that cannot be synthesized by the body. It can be converted to the amino acid *tyrosine,* which is essential for a number of important body functions including the synthesis of the hormones epinephrine (adrenaline) and norepinephrine. The primary source of tyrosine is from the conversion of phenyl alanine. Phenyl alanine is *not* essential if tyrosine is included in the diet (see *phenyl ketonuria*).

Phenyl ketonuria (PKU): A genetically inherited metabolic error affecting the enzyme responsible for converting the amino acid phenyl alanine to tyrosine. Large quantities of phenyl alanine build up in the body and damage the developing brain of newborns. The disease can be treated by removing phenyl alanine from the diet during the critical years of brain growth and development.

Pleiotropic mutation: A mutation affecting several different traits.

Polymer: A very large molecule formed by joining many identical or very similar small molecules. For example, glycogen is a polymer formed from joining many glucose molecules.

Polysaccharide: A biological polymer composed of sugar subunits, e.g., starch or cellulose.

Protein: Any of many different kinds of macromolecules that contain one or more chains of linked amino acids in a definite sequence.

Recessive allele: An allele whose phenotypic effect is not expressed in the heterozygote.

Replication: DNA synthesis; a process in which one DNA molecule serves as a template to produce two identical copies.

Ribosome: A complex intracellular structure involved in the translation of messenger RNA into a protein. It consists of rRNA and proteins.

RNA (ribonucleic acid): A single-stranded nucleic acid similar to DNA but having ribose sugar rather than deoxyribose sugar attached to the bases.

mRNA (messenger RNA): An RNA molecule transcribed (copied) from the DNA of a gene and from which a protein is translated by the action of ribosomes and transfer RNA.

rRNA (ribosomal RNA): A special RNA found in ribosomes. It is *not* translated into proteins.

tRNA (transfer RNA): Small RNA molecules that carry specific amino acids to the ribosome during translation. The tRNA pairs with a homologous place on the mRNA being translated. The tRNA molecule then inserts the amino acid into the growing protein chain.

Sex chromosome: A chromosome correlating with sex; in humans they are the X and Y chromosomes.

Sex-limited: The expression of an inherited trait in only one sex, e.g., baldness in men.

Sex linkage: The presence of a gene on a sex chromosome.

Sexual selection: The differential ability of individuals of different genetic types to acquire mates.

Standard deviation: The square root of the variance.

Statistical significance: A convention by which one determines whether an average difference in measurements between two groups is "real" or has occurred by chance. One uses a statistical test that takes into account the size of the sample and the amount of variation within each sample. Differences that have less than a 5 percent chance of having occurred by chance are considered real and are called statistically significant.

Steroid: See *Hormones (steroid)*

Sympathetic nervous system: Part of the autonomic nervous system that supplies with nerves the viscera, blood vessels, skin glands, and other muscles under involuntary control. Impulses from the sympathetic nervous system generally stimulate the activity of the muscles they supply with nerves; those from the *parasympathetic system* generally depress muscular activity. Involuntary muscles are thus controlled by a balance between the sympathetic and parasympathetic systems.

Testosterone: The principle androgen secreted by the testes; important during embryonic development in the conversion of the indifferent gonad to a testis.

Transcription: The synthesis of RNA using a DNA template.

Translation: The ribosome-mediated production of a protein whose amino acid sequence is derived from the messenger RNA template.

Turner's Syndrome: An abnormal phenotype of the human female produced by the presence of only one X chromosome (XO).

Tyrosine: See *phenyl alanine.*

Variance: A measure of the variation around the central class of a distribution; the average squared deviation of the observations from their mean value.

X-linkage: The presence of a gene on the X chromosome but not on the Y.

XO: See *Turner's syndrome.*

INDEX

249

INDEX

255